S0-EGN-264

WITHDRAWN
UTSA Libraries

▶ **Philosophies of Environmental Education and Democracy**

DOI: 10.1057/9781137484215.0001

The Cultural and Social Foundations of Education

Series Editor: A. G. Rud, Distinguished Professor in the College of Education of Washington State University, USA.

The Palgrave Pivot series on the Cultural and Social Foundations of Education seeks to understand educational practices around the world through the interpretive lenses provided by the disciplines of philosophy, history, sociology, politics, and cultural studies. This series focuses on the following major themes: democracy and social justice, ethics, sustainability education, technology, and the imagination. It publishes the best current thinking on those topics, as well as reconsideration of historical figures and major thinkers in education.

Titles include:

Craig A. Cunningham
SYSTEMS THEORY FOR PRAGMATIC SCHOOLING
Toward Principles of Democratic Education

Aaron Stoller
KNOWING AND LEARNING AS CREATIVE ACTION
A Reexamination of the Epistemological Foundations of Education

Sue Ellen Henry
CHILDREN'S BODIES IN SCHOOLS
Corporeal Performances of Social Class

Clarence W. Joldersma
A LEVINASIAN ETHICS FOR EDUCATION'S COMMONPLACES
Between Calling and Inspiration

Joseph Watras
PHILOSOPHIES OF ENVIRONMENTAL EDUCATION AND DEMOCRACY
Harris, Dewey, and Bateson on Human Freedoms in Nature

DOI: 10.1057/9781137484215.0001

palgrave▸pivot

Philosophies of Environmental Education and Democracy: Harris, Dewey, and Bateson on Human Freedoms in Nature

Joseph Watras
Professor, University of Dayton, USA

palgrave
macmillan

DOI: 10.1057/9781137484215.0001

PHILOSOPHIES OF ENVIRONMENTAL EDUCATION AND DEMOCRACY
Copyright © Joseph Watras, 2015.

All rights reserved.

First published in 2015 by
PALGRAVE MACMILLAN*
in the United States—a division of St. Martin's Press LLC,
175 Fifth Avenue, New York, NY 10010.

Where this book is distributed in the UK, Europe and the rest of the world, this is by Palgrave Macmillan, a division of Macmillan Publishers Limited, registered in England, company number 785998, of Houndmills, Basingstoke, Hampshire RG21 6XS.

Palgrave Macmillan is the global academic imprint of the above companies and has companies and representatives throughout the world.

Palgrave® and Macmillan® are registered trademarks in the United States, the United Kingdom, Europe and other countries.

ISBN: 978–1–137–48423–9 EPUB
ISBN: 978–1–137–48421–5 PDF
ISBN: 978–1–137–48420–8 Hardback

Library of Congress Cataloging-in-Publication Data is available from the Library of Congress.

A catalogue record of the book is available from the British Library.

First edition: 2015

www.palgrave.com/pivot

DOI: 10.1057/9781137484215

Library
University of Texas
at San Antonio

Contents

palgrave▶pivot

www.palgrave.com/pivot

Introduction

Abstract: *An important step toward preserving the natural world is for people to adopt an ethical framework that facilitates such an effort. This is the view of this book. The problems seem to stem from difficulties Alexis de Tocqueville observed about American democracy. He warned that attitudes of materialism, individualism, and conformity could destroy the values people sought, and they encouraged people to abuse the environment. Three philosophers of education, William Torrey Harris, John Dewey, and Gregory Bateson, suggested ways to construct a set of ethics that would offset those flaws.*

Keywords: democracy; ecology; environmental education

Watras, Joseph. *Philosophies of Environmental Education and Democracy: Harris, Dewey, and Bateson on Human Freedoms in Nature.* New York: Palgrave Macmillan, 2015. DOI: 10.1057/9781137484215.0002.

This book offers a unique approach to a serious problem. The problem is environmental destruction. The approach is unique because the book does not recommend any specific actions such as the reduction of the population, the initiation of a public campaign to end capitalism, or the return to sustainable practices of indigenous peoples. Instead, the book contends that the problems of environmental destruction derive from misunderstandings in the popular views of democracy. Accordingly, the way to environmental protection lies in correcting those misconceptions. Although the suggestions to improve the ideal of democracy are not new, the uniqueness of the book is in the idea that advancing democratic ideas may serve the cause of environmental protection as well.

The book is based on the arguments that Gordon G. Whitney advanced in his book *From Coastal Wilderness to Fruited Plain*. According to Whitney, the environmental situation is not desperate even though it is serious. North America has not lost all its natural resources although human economic activity had destroyed several ecosystems. Fortunately, careful management revived some species of plants and animals. For Whitney, the important element is whether people can manage the ecosystem in ways that work within limits of what the land can allow. This is essential, he argued, because America does not have unlimited supplies of water, soil, plants, animals, or minerals. Fortunately, environmental protection is not difficult. Whitney claimed that the actions are simple. The important element that requires the most effort is a change in thinking. According to Whitney, if Americans are to protect their environment, they must adopt an ethical framework that advances environmentally sensitive ways of living and working.[1]

This book suggests that such a framework need not be new or foreign to Americans. It looks at three intellectuals who developed ideas that Americans could have about the nature of a good education and appropriate human relations. These philosophers established ways of thinking that fit within the patterns of American democracy. For this reason, they offered perspectives that served as alternatives appropriate to the American culture. Although they encouraged people to change their conceptions about their relation to the environment, they did so by introducing ways of thinking that would have many beneficial effects.

During the period following the U.S. Civil War, the country grew and cities built school systems. William Torrey Harris offered a curriculum that showed students how social restraints could enhance everyone's freedoms. He arranged John Locke's theory of private property to require owners to use the things they control in socially beneficial ways. During

DOI: 10.1057/9781137484215.0002

the Progressive Movement, John Dewey followed Harris in rejecting dualisms that separated individuals from society, and he went beyond Harris to claim that the pursuit of material rewards for labor cheapened the value of work and led to environmental destruction. With the advent of postmodernism, Gregory Bateson expanded Harris's concern for society to suggest that all things on earth were related and schools should show how efforts to shape the environment to fit human desires would lead to the destruction of the environment and of the people who depended on it.

As the second chapter will show, many scholars consider the concerns for environmental protection to be a new political consciousness. The conservation movement as a political effort gathered force during the presidential administration of Theodore Roosevelt; however, some scholars argue that the belief did not arise until the 1970s that any drastic change in the ecological system was wrong or harmful. While those scholars may be correct, this book will contend that the foundation for an ethical stance in favor of preserving the ecology began in the United States much earlier. This is because the problems of environmental destruction derived from the difficulties that Alexis de Tocqueville found in the ways Americans interpreted the ideal of democracy. At least, this is the point of view the book takes.

In the 1830s, Tocqueville noticed that Americans had three tendencies that turned them away from the ethical orientations that democracy made possible. Since democracy implied rule by the people, it offered the chances for people to form communities in which they shared in the good things of life. Unfortunately, the drive for materialism, the appearance of a new form of individualism, and the tendency toward conformity threatened the possibility of democracy remaining true to its promise. This book will argue that those same tendencies led Americans to act badly toward Native Americans and to abuse the environment.

At the same time, though, this book will discuss the ideas of Harris, Dewey, and Bateson specifically because they offered ways to offset the difficulties of materialism, individualism, and conformity. There were other writers or scholars who did the same. In fact, some of the figures in this book looked to these other scholars for inspiration. For example, Dewey acknowledged a debt to Emerson in an essay he wrote in 1903.[2]

Harris, Dewey, and Bateson serve as important figures for this book, in part, because they blended their understandings of the dangers of American versions of democracy with suggestions of how schools could improve people's ways of thinking. Furthermore, their ideas fit together

DOI: 10.1057/9781137484215.0002

in a sort of progression so that each suggestion illuminated the previous one to some extent.

In Chapter 1, readers will encounter a brief description of the development of an ecological ethic and how scholars thought Americans would have to reverse the traditional views they had about property if they were to fulfill their responsibilities to the land on which they built their society. It is in this chapter that readers will meet the contention of Alexis de Tocqueville that the attitudes Americans developed from their version of democracy were corrosive.

The subsequent chapters are devoted to the ways each of the philosophers suggested reducing the problems of materialism, individualism, and conformity. For example, Chapter 2 considers the ideas of Harris. Chapter 3 looks at Dewey's contributions, and Chapter 4 describes the ways Bateson might broaden the American desire to solve the world's problems. Finally, Chapter 5 will review how the philosophers offer ways of thinking that would broaden people's views of democracy and reduce the problems of environmental destruction. These were alternatives that teachers could introduce in schools, and they might serve as a basis for an ecological ethic. One difficulty this conclusion describes is that many commentators advocate solutions that extend the patterns of thinking behind the ecological difficulties. For this reason, those solutions might make the situation worse.

It is reasonable for the suggestions to consider what school people can do because this book follows Whitney's idea that people have to develop an ethical framework if they wish to save the natural environment. Schools should be a place where people learn to think in humane and beneficial ways; however, for this to happen, educators may have to reconsider the aim of education in Western societies. It might mean that schools will devote less attention preparing children for the world of work and spend more time showing how limiting economic resources can improve the quality of life.

Notes

1   Gordon G. Whitney, *From Coastal Wilderness to Fruited Plain: A History of Environmental Change in Temperate North America 1500 to the Present* (Cambridge, UK: Cambridge University Press, 1994), 337.

2   John Dewey, "Emerson—The Philosopher of Democracy," *International Journal of Ethics*, vol. 13, no. 4 (July 1903): 405–413, Stable URL: http://www.jstor.org/stable/2376270, accessed 31 December 2014.

DOI: 10.1057/9781137484215.0002

1
Defining the Task

Abstract: *Settlers brought from Britain to the eastern shores of the United States several ideas of property and democracy that caused important problems. In the 1830s, Alexis de Tocqueville noted three tendencies among the settlers that threatened to destroy the benefits of democracy and the environment. They were individualism, materialism, and conformity. By the 1970s, the concept of ecology changed from a scientific term into a moral critique that urged people to restrain those tendencies. The process was difficult. Some critics argued that democratic governments were less effective in enacting conservation policies than authoritarian ones.*

Keywords: Alexis de Tocqueville; John Locke; national parks

Watras, Joseph. *Philosophies of Environmental Education and Democracy: Harris, Dewey, and Bateson on Human Freedoms in Nature.* New York: Palgrave Macmillan, 2015. DOI: 10.1057/9781137484215.0003.

In recent years, colleges and universities around the world have intro-duced programs in environmental education. According to David John Frank, Karen Jeong Robinson, and Jared Olesen, such offerings appeared in the curriculums of universities around the world in the 1970s, and they spread dramatically in the 1990s. Such programs became the fastest growing academic area in the United States. At the same time, employers significantly increased the opportunities for graduates of such programs. According to Frank, Robinson, and Olesen, the growth of these univer-sity programs was not related to local needs but to a set of views that diffused rapidly and widely.[1]

Views of environmental sensitivity, sustainability, or ecological aware-ness may be a new perspective that spread to university campuses, as Frank, Robinson, and Olesen claim; however, the foundation for such a set of concerns began earlier. According to Anna Bramwell, the develop-ment of ecological awareness as an ethical position was a new political consciousness that derived from two strands. One strand began in the last quarter of the nineteenth century, and it built on the work of German zoologist Ernst Haekel, who coined the word "ecology" in 1866 to indicate the study of the relations among organisms and their environments. The second strand was an economics of energy that focused on the problem of non-renewable resources. She argued these ideas fused in the 1970s to produce an intensely conservative, moral critique that became popular when the twentieth century ended. Bramwell defined the ethical view of ecology as the recognition that energy flows within a closed system and that any drastic change within the system would be wrong or harmful.[2]

Although Bramwell focused her history on changes in the United Kingdom and Europe, a survey of articles in academic journals illustrated that the same strands appeared among scientists and literary authors in the United States. At the turn of the century, scientists used the term "ecology" to turn scientific research from constructing abstract theories to the discovery of connections among living things that could improve human agriculture and industry. By the 1970s, humanists borrowed the term as a call to unite people in a movement to restore democracy and cooperation under a traditional pastoral ideal. The following section may make clear this pattern of development.

Development of the field of ecology in America

In America, scientists began using the term "ecology" to define research that uncovered methods to improve human society. For example, when

DOI: 10.1057/9781137484215.0003

V. M. Spalding delivered his presidential address to the 1902 meeting of the Society for Plant Morphology and Physiology, he explained to his listeners that the term had entered the vocabularies of scientists only a few years earlier to define an academic area that he hoped would show how to rejuvenate forests and enhance agricultural production. In his address, Spalding noted that colleges in the United States had changed the instruction of biology and botany within a period of about 25 years. He pointed to a dramatic increase in the number of courses, books, and journals covering the biological sciences that were especially devoted to ecology to the extent that the word "ecology" had become rooted in university studies. According to Spalding, the person who advanced the work of ecologists before the field had a name was Charles Darwin, because he sought the origins of living forms by studying them in the conditions within which they lived. Noting that subsequent researchers applied the tools previously unconnected to botany, such as statistics, to determine the ways nature changed, Spalding urged the scientists in his audience to move quickly to undertake disciplined research that could apply findings about ecology to fields such as forestry and agriculture.[3]

The president of the Ecological Society of America made a similar plea in 1921 to include human activities in the study of ecology. Acknowledging that including human activities would make the field of ecology into an applied area of study, Stephen A. Forbes noted that the state of Illinois employed an ecologist and an entomologist to determine the ways weather conditions influenced the codling moth's life cycle. This moth destroyed fruit tree crops, and insecticides worked best when they caught the larvae at particular stages of development. Since weather conditions affected the maturation of the moths, this information was valuable to farmers. Forbes termed such an application of science to farming as "the humanization of ecology."[4]

As late as 1957, ecologists called on their colleagues to make careful, objective experiments that could clarify the aims of the field. Writing in *Ecology*, the journal of the Ecological Society of America, Richard S. Miller complained that the standard definition of "ecology" as the study of the relations between an organism and its environment did not distinguish this field from other biological sciences. He suggested that ecologists should recognize that they seek to discover the biological properties of populations and communities. He added that locating the properties of populations would separate the work of ecologists from those of biologists because the latter focused on individual organisms. More important, Miller thought that ecologists should conduct

DOI: 10.1057/9781137484215.0003

experiments producing measurements that defined those characteristics. It was not enough, he argued, to make observations of organisms in their environments.[5]

By 1970, the popular use of the term "ecology" included the notion that the spread of environmental destruction threatened the continued existence of human beings. For example, Leo Marx published an essay in the journal *Science* noting that competing groups expressed concerns about the preservation of nature. He suggested that since federal and state governments had rarely tried to preserve natural areas, voluntary organizations had sought to preserve the outdoor life they wanted to enjoy. Affluent and interested Americans joined organizations such as the Sierra Club or the National Wildlife Federation to work outside city limits; however, Marx noted they seldom tied their efforts to the welfare of people living in poverty. The advent of the Cold War encouraged younger people who disliked American middle-class life to argue that atomic destruction and chemical pollution threatened everyone's lives. Marx argued that these dissenters fused a contemporary argument with a long-standing American image of the benefits of a pastoral life to forge an ecological movement.[6]

Although Marx claimed the inspiration for the ecological movement came from utopian nature writers, he thought that scientists from the American Association for the Advancement of Science could help resolve the problems. Marx recommended a three point strategy. The scientists could form a panel that met regularly to investigate critical environmental problems, evaluate the efficacy of government efforts up to that point, and suggest remedies. He wanted environmentalists to join forces to change people's conceptions of the types of lives they should live, and pushed for a campaign to influence public and private institutions to limit the pursuit of profits to allow for practices that would preserve the environment.[7]

In the 1970s, educators began to consider ways schools could influence people to reduce the destruction of the environment. For example, James Wheeler and Nobuo Shimahara considered what they called the ethical aspects of the ecological crisis. They listed ten illusions about the economic and social conditions that led people to contribute to environmental problems. These included the belief that people had to own the latest automobile, a fine house, and the best clothes. The belief that full employment meant the economy had to grow continually required increasing amounts of natural resources. They suggested that schools

DOI: 10.1057/9781137484215.0003

could counter these misapprehensions with a growing recognition that the integrity of nature had to be protected and that people's desires had to reflect this obligation.[8]

Despite the humanists' and educators' pleas for public campaigns and educational programs, political scientists argued that environmental problems could not be solved easily. According to Susan M. Leeson, the traditional model of property in America came from John Locke. His view was that people created private property by working on a piece of land to bring some provisions for themselves or for others. She added that the availability of land in the American frontier seemed to make this view the basis of American political thought even though critics noted that there was not enough land for everyone. As the open or available land disappeared, the National Environmental Policy Act of 1970 required federal proposals to state what would be the project's impact on the environment. Despite this legislation, Leeson noted that federal agencies submitted proposals after 1970 for projects such as urban highway construction without revealing concern for natural resources. According to Leeson, several political scientists believed the only solution to environmental problems was for the government to become severely authoritarian. Leeson concluded that the ecological crisis offered an opportunity for people to choose the society they wanted. If people wanted to live in a democratic society, they would have to forgo their desires for cars, houses, and clothes. If they continued to circumvent reasonable regulations, some forceful agency would have to control their animal drives for human life to continue. Leeson hoped that people could reaffirm their capacities for reason and create an intelligible and sustainable order for themselves.[9]

Traditional American views of relation of society to the environment

When Leeson claimed John Locke provided the basis of American's views of the relationship of nature and personal property, she repeated the ideas of many political scientists that Locke's treatises of government principles underlay the creation of modern democracies. Locke divided the treatises into two parts. In the first, he refuted the view that monarchs had a right to rule. In the second, he argued that people justified their ownership of property by laboring upon the land to bring forth goods that people could use.[10] This was the idea that Harris used.

DOI: 10.1057/9781137484215.0003

Writing in 1689, Locke argued that God had given the world and all its creatures to humankind for them to use for their comfort and convenience. Although this implied that nature's bounty was for all people, Locke used America as an example to contend that the fruit or venison that an Indian consumed belonged to that Indian because it had become part of his or her body. Since the labor of any person belonged to the laborer, Locke extended this idea to contend that people came to possess some part of nature when they mixed their labor with it. To place reasonable limits on this perspective, he cautioned that individuals could possess the rights to a piece of property provided that there was enough remaining for everyone else. For example, water running in a fountain may be held in common, but when people filled their pitchers with some, the portion in the vessels became their property. In the same way, Locke contended that the deer an Indian slew belonged to that hunter. In both cases, water remained in the fountain for other people and there were more deer in the forest to satisfy other hunters.[11]

In placing limits on the possession of property, Locke tried to contain the effects of human greed. On the one hand, he argued that it was unjust for a person to take what another person had previously improved with his or her labor. On the other hand, he contended that a person should not acquire more of anything than he or she could reasonably use because wasting resources injured other people. Waste denied other people resources they could have used. He did not consider it a danger for people to cultivate large estates provided the owner did not allow the excess produce to go to waste. Locke believed there was enough land for everyone to use, and the products of cultivated land could improve the conditions of many people.[12]

Although Leeson may be correct in assuming that Locke's ideal of private property dominated American political thought, the federal government resisted the dispersion of public lands for several years. According to Peter Onuf, the Land Ordinance of 1785 required survey teams to divide open territories into townships, but this was not to prepare for settlement. Onuf added that the surveys were to prevent individuals from wandering into an area and claiming sections of land for themselves.[13]

The Continental Congress that met from 1774 to 1789, and the U.S. Congress that followed, sought to control immigration into the frontier lands. The Northwest Ordinance of 1787 went further. It forbade settlers from taking Native American lands without permission specifying that

DOI: 10.1057/9781137484215.0003

the Native Americans had property rights, and it prohibited settlers from harming the Native Americans except during lawful war declared by the U.S. Congress. These controls restrained immigration to the point that only one state, Ohio, gained enough population to enter the Union until the War of 1812 ended. During the four years after the war, immigration into frontier land increased to the point where four states entered the Union. Although some settlers had taken land illegally, the government permitted those settlers to buy their portions. Other settlers could take possession with no more than a promise to pay. Although the U.S. presidents had tried to protect frontier lands, they relented in the face of the resulting protests. Their policies secured the spread of the population into the frontier lands occupied by Native Americans.[14]

Believing that Americans should spread into the frontier, U.S. President Andrew Jackson twisted Locke's views to justify the disinheritance of the Cherokee Nation from their lands in Georgia. The issue was controversial because several American organizations opposed Jackson's efforts in this case.

One set of opinions that opposed Jackson came from the U.S. Supreme Court whose justices had decided in three cases between 1823 and 1832 that the Cherokee Nation had the right to occupy the lands on which they had settled as firmly as white people do when they purchase a parcel of land. According to Angela R. Riley, the problem was that white settlers wanted the land, and the federal government decided it could not protect the Native Americans. In 1830, Jackson addressed Congress to justify the displacement of the Cherokee. He claimed it was better to turn the expanse of land over to millions of farmers who would work their own fields than it was to allow a few thousand Native Americans to hunt in the forests as they pleased. The Congress agreed and passed a series of Removal Acts that led to the dispossession of the Cherokee.[15]

Although Jackson repeated Locke's notion that private property depended on labor, the American Board of Commissioners for Foreign Missions (ABCFM) found that Jackson misrepresented the conditions in which the Cherokee lived. This was the second group that opposed Jackson's efforts. The commissioners represented the Cherokee in the U.S. Senate. ABCFM representatives argued that the federal government had entered into treaties agreeing to protect the Native Americans' rights to remain on their lands. They pointed out that U.S. President James Monroe had visited Georgia in 1819 to witness the progress of the missionary efforts and been impressed enough to lead the Cherokee to believe they would not be removed. By 1821, the Cherokee developed a written language and established newspapers. Nonetheless, members of the Georgia legislature pressed the U.S. Congress to give the Cherokee lands to the state.[16]

After Jackson made his speech to Congress, the Georgia legislature annexed Cherokee land. The U.S. Congress began considering the issue and received a bill requesting approval of the actions of the Georgia legislature. In reply, the ABCFM asked Congress to delay voting on the bill until an investigating committee could visit the Cherokee in Georgia to ascertain the situation to verify Jackson's accusations about their state of civilization. The vote to delay the decision until there had been an investigation failed by one vote. The vote to remove the Native Americans barely passed, but in 1838, the forced removal of the Cherokee and the subsequent trail of tears began.[17]

Tocqueville's analysis of democracy in America

Alexis de Tocqueville visited the United States for nine months from 1831 to 1832 when the controversy over the Cherokee lands was under way. Although Tocqueville did not comment on Jackson's actions, he observed several tendencies in the American character that seemed to encourage Americans to act badly toward Native Americans. Publishing his observations in 1835 and 1840, Tocqueville argued that tendencies toward materialism, individualism, and conformity characterized the society. Ironically, these tendencies came from the conditions of American democracy, yet they threatened democracy itself. This happened because the tendencies led people to ignore wider ethical orientations.

At the risk of over-generalizing, the way the Americans treated the Native Americans illustrated one aspect of the American tendency to distort the ideal of democracy. To justify the dislocation of the Cherokee, Jackson appealed to an American wish to conquer the land and make it into European settlements. In this way, Jackson's actions served the tendency of Americans to covet material goods. Since the dispossession came at a high human price, the action showed that Americans lacked ethical boundaries that could limit their materialism. Accordingly, they could abuse the environment they had occupied. Fortunately, as the earlier discussion shows, the American population did not unanimously support Jackson's actions, thereby leaving some possibilities for developing those limits.

DOI: 10.1057/9781137484215.0003

When Tocqueville described the miseries he witnessed the Native Americans suffering as they migrated to new lands before the U.S. government forced the removal of the Cherokee, he described the problems as deriving from a type of selfishness. At that time, he wrote that white Europeans did not force the Native Americans to travel but hunger did. In the places where Tocqueville watched the process, white Europeans caused the hunger that drove the Native Americans west. A few European families would move into an area and the animals would flee. The Native Americans had to follow the game since they depended on those animals for food. As Europeans advanced, the United States government made treaties offering to pay for the land in whiskey, clothing, and firearms. According to Tocqueville, such trades reduced the Native Americans to penury because they developed desires to have manufactured goods but they had lost the ability to hunt for the animal furs to buy those goods. In this way, Tocqueville claimed the line of white settlements moved inexorably west pushing the Native Americans further and further away.[18]

According to Tocqueville, the desire for material gain was pervasive among Americans. The poor people in America never surrendered the longing for rewards they never received. Many rich people, who had been born in circumstances of poverty, were not satisfied with their newly acquired wealth. They longed for more. Tocqueville thought such longings were attributes of the middle class. These people were absorbed in their efforts to accumulate material goods because they feared those goods could slip away. Since most Americans were neither very poor nor very rich, many people adopted this grasping nature. Tocqueville thought that it caused Americans to be continually unsatisfied and always moving on to find new but eventually unsatisfying material pleasures.[19]

Individualism was the second tendency that Tocqueville thought arose from democracy and threatened it. According to him, individualism had developed recently in the eighteenth century. It differed from egoism, which he considered to be an exaggerated love of self. Individualism was the tendency for a person to pull away from society and find security within a small circle of friends and family. Ironically, Tocqueville thought it originated from the spread of equality in American society. The reasoning was as follows: Americans confused equality with liberty; they thought that everyone was free because they were equal. They made this mistake because, when conditions were equal, people could not assert privileges over other people or demand deference because of status.

DOI: 10.1057/9781137484215.0003

Most important, equality was easy to protect while liberty demanded continued vigilance in the arena of politics because complex changes in laws could restrict some people and advance others. Tocqueville argued that few Americans recognized the problems and maintained equality and the individualism that came from it. Since everyone was equal there were no ties among members of communities. People did not expect anything of anyone else and did not feel obliged to help anyone.[20]

Such a pattern of beliefs made it easy for Americans to overlook the restraints that Locke had placed on his theory of property. After all, Locke did not prevent people from taking large estates. He justified such acquisitiveness by recognizing the production of food would benefit society. Nonetheless, he did warn that people should not take what other people had created. The missionaries argued that the Cherokee had settled the land, but when the European settlers arrived they claimed they discovered free open areas they could possess. When the Native Americans complained to the courts and to the U.S. Congress, the authorities did not attend to their views.

The third tendency that distorted democracy was conformity, and this may explain why authorities did not attend to the Native American petitions. Once people felt that all the land was for white settlers, few people could disagree. In fact, Tocqueville noted that Americans ignored their freedoms to accommodate public opinion. Tocqueville traced the dominance of the majority to the democratic government. In part, the majority had such authority because people tended to believe that there was more wisdom in a group than there was in an individual. This idea spread slowly, Tocqueville added, but it seemed irresistible. He gave an example of mob violence against a newspaper that published editorials criticizing the War of 1812, which at the time was popular. Neither the city officials nor the militia could protect the lives of the journalists from the fury of the crowd. In another instance, Tocqueville asked a citizen why free African Americans did not exercise their right to vote in Pennsylvania. The answer he received was that the black people chose not to visit the polls to cast their votes. Although they had the legal right to vote, popular prejudice stood against the practice and the African Americans did not contradict public opinion. From examples such as these, Tocqueville decided that majority opinion exercised a more severe tyranny than did any law.[21]

Conformity could allow Americans to mistreat Native Americans because once a prejudice was in place rational argument could not dislodge it. The votes in Congress for taking the Cherokee lands in Georgia illustrated that this course of action was widely approved. Many people felt that American farmers should have the land that belonged to Native Americans. In addition, the examples of mob violence against newspapers during the War of 1812 illustrated the emotional power of groups to overshadow ethical or legal restraints.

The importance of Tocqueville's ideas may be illustrated by the fact that they retained relevance into the twenty-first century.[22] For this essay, it is more significant that they received careful attention on their publication. For example, John Stuart Mill wrote a most appreciative and lengthy review of *Democracy in America*. Mill praised Tocqueville for applying the method of a naturalist to political philosophy. He wrote that Tocqueville followed scientific principles by considering the political aspects of democracy in America as a reality. Mill noted that Tocqueville took the laws of human nature as people understood them and showed how they unfolded in the social conditions of the new and rapidly growing country. He agreed that Tocqueville accurately described how democracy had released the tendencies of materialism, individualism, and conformity. The point on which Mill disagreed was the question of whether these tendencies derived from the development of democracy or from the unique conditions of America.[23]

Mill built his criticism on an admission Tocqueville made in his book. In explaining why he came to the United States, Tocqueville wrote that he did not come to look at this country in particular. To him, the United States provided a test of what would happen when the features of democratic society swept across the world. He offered a brief view of historical developments that suggested the shift from aristocratic privilege to equality of condition in Western civilization was inexorable. More important, he noted that when the first immigrants came to America from England, the only system of government that they brought to the United States was democratic. Thus, the ideal of equality grew in those former colonies unfettered by competing views.[24]

Although Mill agreed with everything that Tocqueville wrote in that introduction, he thought Tocqueville had mistaken the influence of democracy with the effects of commercialism. Mill praised Tocqueville for recognizing that democracy was not a form of government but a social condition wherein everyone was equal. This distinguished England from America. Mill thought English society contained three classes. The commercial class was becoming dominant, but the learned class, the leisured class, and the agricultural class remained to counter-balance the dangerous tendencies of commercialism. The problem in America was that everyone was part of the commercial class and was devoted to material gain.[25]

Mill claimed the tendencies of the agricultural class opposed and thereby balanced the tendencies of the commercially oriented one. According to him, farmers in England did not try to become rich although the farms enjoyed sufficient profit to maintain themselves. The agriculturalists enjoyed their occupations and took pleasure from the work they did. In fact, several people looked to retire to become farmers as a form of amusement. Mill thought this class had to remain as they were to keep alive the moderate wishes and tranquil tastes that fit their life on farms. The other two classes offered similar balance from the drive of the commercial class, and together with the agricultural class, they would prevent the English from becoming as grasping and conformist as the Americans.[26]

How materialism, individualism, and conformity might encourage environmental abuse

In part, the apparently boundless landscape may have prompted dreams of material success and, when distorted, encouraged selfishness. In 1782, when J. Hector St. John de Crèvecoeur described what was unique about America, he noted it was the liberty that people enjoyed and the substances they possessed. The rich and the poor people did not stay apart because there was no aristocracy. He noted that in the new land everybody worked for himself or herself. They came from many different countries, yet they blended into one race. According to Crèvecoeur, a man arrived as a European; however, he soon learned that he could work for someone and acquire land for himself provided he was sober, industrious, and honest. It was the feeling of ownership that caused many of the newly arrived immigrants to feel independent and self-reliant.[27]

Several years after Crèvecoeur penned his letter, Tocqueville made a similar observation about boundless riches in the United States. According to Tocqueville, the northern shores of America appeared ominous; however, inside the external boundary there was a thick forest where life flourished. More important, Tocqueville thought the vegetation or the animals were unharmed by destructive forces. Vines emerged

DOI: 10.1057/9781137484215.0003

among dead trees, broke up the wood, and made spaces for new sprouts. To Tocqueville, the plentiful plant and animal life suggested that the land would serve as the place for a great nation to appear. The Atlantic coasts invited trade and industry; the rivers in the Mississippi valley seemed to flow endlessly.[28]

About a year after Tocqueville made his observations, Ralph Waldo Emerson published his first book entitled *Nature.* Writing for readers in England, Emerson expressed a more religious explanation of why America's natural resources awaited human hands. He noted that all parts of nature advanced the profits of human beings. Wind sowed seeds. Sun and wind cooperated to turn the seas into rain that fed the plants, which provided fodder for animals. Human beings nourished themselves from those beasts. These miracles enabled human beings to work producing reproductions of nature with steam, railroads, and canals. Emerson concluded that these material benefits were one of four ways that nature worked with human beings to improve their conditions. The other ways included Beauty, Language, and Discipline. Accordingly, Emerson considered nature to be a force that God provided to human beings that they could use to improve their culture.[29]

In his essay, Emerson did not invite human beings to plunder nature. Instead, he expressed his faith that God created nature to help human beings if they were careful enough to recognize how to use it. In this regard, Emerson offered a guarded view of the influence of the bounty of America, and the extent of nature's gifts humbled him. On the other hand, the riches Crèvecoeur surveyed on his farm gave him a sense of self-satisfaction.

Henry David Thoreau noticed the apparent inexhaustibility of nature in 1854 as he walked home from fishing near a bridge in Concord, Massachusetts. While Thoreau looked at the birds, flowers, and trees, he decided that Nature was so full of life that the death of many animals or plants passed without notice. Herons ate tadpoles, the wheels of wagons killed toads, and life went on despite these tragedies. In fact, Thoreau considered these disasters to be beneficial. For example, the heron gained health and strength from the tadpoles it ate.[30]

In making such observations, Thoreau was not being callous. He was sensitive to the beauties of nature, and he recognized the harm that people caused themselves when they greedily accumulated possessions. Most important, he was a systematic and careful naturalist. His observations of the blooming dates of native plants were so accurate that

DOI: 10.1057/9781137484215.0003

scientists today use them as base lines for measuring the advances of global warming.[31]

According to Donald Worster, Thoreau showed his neighbors how they could manage the forests by observing the animals around the city of Concord. New Englanders were proving to Thoreau that nature's bounty had limits. Concord had supported a dense forest in 1638 when it was settled by English Puritans. The trees included hemlocks, chestnuts, and maples. The grandest specie was white pine. According to Worster, white pine trees grew to a height of 250 feet with a width of 6 feet. By 1700, however, the residents had cleared the trees and removed the forest. The event that prompted Thoreau's concern came from George B. Emerson, president of Boston's Society of Natural History. Writing a report on Massachusetts's forests in 1846, Emerson complained that people cut all the trees in each area of the woods they entered. Acknowledging that people used the parts of the trees for homes, wagons, ships, and household items as well as for fuel, he predicted that the practice of clear cutting would destroy this valuable resource.[32]

Thoreau offered a solution by recalling his days at Walden where he watched squirrels bury hickory nuts and a tree sprouted shortly afterward. He turned his observations into a lecture entitled "The Succession of Forest Trees," in which he recommended that farmers plant trees in the natural order of succession. This meant that pines came first, and oaks would follow. Although Thoreau was appalled at the way people ignored or abused nature, Worster praised Thoreau for setting an example and acting like an ecologist by determining what nature did to protect itself.[33]

Although Thoreau offered a reasonable way to preserve the forests, the destruction of forests reached a fever pitch before the twentieth century began. Gifford Pinchot is credited with beginning the conservation movement; he was the first chief of the U.S. Forest Service. According to Pinchot, the prevailing view in the government and among the members of the public until the first years of the twentieth century was that the frontier was a place to be settled. Pinchot wrote that although colonial immigrants brought from England a view that forests had to be preserved, they changed this view when they realized the forests helped Native Americans to attack the colonists and that tall trees prevented growing crops. The view that forests hindered progress prevailed in the United States and this encouraged people to exploit the wealth that they found in timber. Quoting a report from the New York State Forest

DOI: 10.1057/9781137484215.0003

Commission published in 1886, Pinchot pointed out that it was impossible to apply European models of forest science in the United States. One reason was that the greatest fury of destruction of forests in the world took place in the years leading to the twentieth century in the United States. He believed that it derived from the desire to turn natural resources into money. He thought most people wanted to be rich, and they saw the exploitation of valuable natural resources as a way to realize that goal. Unfortunately, the federal officials who oversaw public lands abetted the process because they thought their duty was to distribute the lands to American citizens who wanted them.[34]

Although there was extensive fraud, considerable waste, and obscene profit taking, Pinchot approved of some legislation that distributed land to settlers, such as the Homestead Law that Abraham Lincoln approved. He called other legislation, such as the Timber and Stone Act of 1878, thoroughly bad. These laws gave the U.S. Secretaries of the Interior and the Commissioners of the General Land Office the idea that they should get rid of the timber that had covered the United States.[35]

To illustrate the abuses, Pinchot recounted an example of a lumber company that gave train-loads of teachers from the Midwest vacations to the redwood lands of California. A company representative took each teacher into the forest and told the teacher that he or she owned that particular quarter section. Under the Timber and Stone Act, the teacher could use the land and the magnificent trees as he or she pleased. The teacher deeded the land to the timber company that paid $2.50 per acre for what Pinchot called the most valuable timber land on earth. After the teachers left, the company harvested the trees.[36]

Pinchot noted that things changed at the federal level when the U.S. Department of Agriculture began in 1898 to hire foresters who could help owners of timber lands manage their holdings to harvest more trees than they would with less carefully selected methods. After Theodore Roosevelt became president in 1901 the foresters joined together in the U.S. Department of the Interior where they exerted some control over the public lands. Pinchot had conceived of starting a unified conservation effort while riding a horse in a forest. As an expert in forest management, he realized that the problems he confronted were related to water pollution and to mineral exploitation and to soil erosion. When his colleagues suggested to Roosevelt that environmental policies should consider these problems as related to each other, the President accepted the idea and termed it "conservation." This became the label for a coordinated effort

DOI: 10.1057/9781137484215.0003

that enabled the government to address the environmental problems together.[37]

Roosevelt appointed a Commission on National Conservation and charged it with compiling an inventory of all resources. Although the report was published in 1909, it was not widely distributed. The U.S. Congress had decided the commission was illegal, and U.S. President Taft agreed. Taft had followed Roosevelt in office. In defiance, conservation advocates formed the National Conservation Association, which was a private organization, to guide the U.S. Congress to adopt legislation protecting natural resources. The directors included friends and associates of John Dewey, such as Jane Addams, with whom he had worked closely at Hull House.[38]

In the essay quoted here, Pinchot made his argument for reasonable control of the use of timber, land, and water with the example of large businesses fraudulently taking possession of land from the federal government. Another factor in the destruction of the forests was the actions of individuals who were convinced they acted correctly.

An example of such an attitude appears in an account William James offered in his book *Talks to Teachers* written in 1899. In one essay, James described an insight he gained during a train trip through the mountains of North Carolina. He observed to his seat-mate that in each little valley a white settler had cut down any modestly sized tree and burned the stumps of the larger ones. The settler had built a rude cabin and constructed a tall rail fence around the patch of forest he had destroyed. When James asked his fellow traveler what sort of person would make such a mess for his home, the fellow responded that he lived in a similar manner and that neither he nor his neighbors were happy until they placed one of those valleys under cultivation. When James heard this response, he realized that what he saw as the destruction of a beautiful natural setting represented to the mountaineer the result of a moral duty to struggle against the environment and create a safe haven in which to live. From this encounter, James recognized that people held different ideals depending on their circumstances.[39]

When James recognized that the people who expressed this striving for material satisfaction did not think it was evil, he showed how people turned this distortion of democracy into an ideal of life. By recognizing how a person's situation influenced his or her perspective, James implied that people had a moral obligation to avoid imposing their views on everyone else. This observation should have placed James in a difficult

DOI: 10.1057/9781137484215.0003

situation. Destroying the forests prevented people from holding them in common, yet the requirement to appreciate different ideals should have demanded restraint from turning forests into waste land. Unfortunately, in *Talks to Teachers*, James did not extend his argument this far.

In fairness, the damage the mountaineers caused that James described was not permanent. For example, Gordon G. Whitney offered a detailed study of the changes in the ecology of northeastern United States in his book *From Coastal Wilderness to Fruited Plain*. Discussing nineteenth-century farming practices, Whitney criticized the popular story of a poorly educated farmer exploiting fertile land until it was exhausted and moving west. Whitney contended that the situation was more complicated than that story implied. According to him, it was often the case that other farmers bought the so-called exhausted land and turned it into profitable farms. Not only did Whitney contradict the popular theory that nineteenth-century farmers wasted the land and left for virgin soil, he offered the hope that people could cooperate without any formal or tacit agreement in ways that could benefit the land.[40]

In a study of the township of Chelsea, Vermont, from 1784 to 1900, Hal S. Barron agreed with Whitney's account. The common view among historians was that farmers in the last half of the nineteenth century moved west, held larger tracts of land, and used increasingly mechanized methods of farming. He described photographs of teams of 20 horses pulling large combines to illustrate this trend. Barron contended that this picture may have been true for California, but most farms in the North were different. Between 1840 and 1900, the population in Chelsea decreased by more than 40 percent; however, this was not the result of decay and decline. According to Barron, the farmers retained their property, diversified their production, and turned to produce that they could sell in nearby cities. They raised sheep, and they turned to agricultural societies to learn how they could manage in the changing economic and social conditions. Most important, they learned about composting, spreading manure, and other techniques with which they could revive the soil.[41]

Setting aside land for preservation

An important motive for preserving natural settings was to maintain nature's beauty for other people to enjoy. This was the aesthetic motive,

DOI: 10.1057/9781137484215.0003

but it seemed to depend on an authoritarian set of policies. It may be from this experience that Susan Leeson, quoted earlier, found several political scientists who believed the only solution to environmental problems was for the government to become severely authoritarian.

As for the authoritarian stimulus in natural parks, Mark David Spence argued that the wish of permitting people to enjoy nature in a pristine form undergirded the efforts of American preservationists, such as John Muir. They wanted to create places of wilderness where vacationing Americans could enjoy the experience of natural surroundings. This meant the parks had to be uninhabited, and the preservationists considered the Native Americans who lived within those areas as obstacles to their plans. When the U.S. Congress created the National Park Service in 1916, the service took charge of supervising activities on the park land and closely monitored the activities of the Native Americans who used those areas. By 1928, the Park Service limited the number of Native Americans who could remain in their villages on government land.[42]

While the U.S. National Park Service oversaw extensive areas of land, Germany's National Socialist government created one of the most wide-ranging conservation laws in 1935. As might be expected, this law offered an example of conservation policies operating under an authoritarian regime.

Historians disagree whether the Nazi party sought to protect the environment or to use the law to disguise other efforts. For example, Bramwell traced environmentalist policies to the Nazi party in her 1985 book *Blood and Soil: Richard Walther Darré and Hitler's Green Party*. She concluded that the Nazi party had a Green wing. When Franz-Joseph Bruggemeier, Marc Cioc, and Thomas Zeller quoted Bramwell, they contended that most historians disagreed with her statement.[43]

Charles E. Closmann claimed that it was not possible to isolate any motive for the actions of the Nazi party. Many different people belonged to the Nazi party, and they held distinct, often contradictory, goals. Nonetheless, Closmann acknowledged that, when World War II began, Nazi troops entered Poland, and Nazi officials exterminated Polish and Jewish villagers in some rural areas to create a national park. According to Closmann, the aim was to return the forests to their primitive state in order to reclaim the mythic origins of the German state.[44]

The authoritarian nature of the Nazi government enabled Nazi officials to carry out their plans expeditiously. Nonetheless, the Nazi experience suggests that authoritarianism might not protect the environment.

DOI: 10.1057/9781137484215.0003

For example, Bramwell explained that Rudolph Steiner, the founder of Anthroposophy and Waldorf schools, developed a program of environmentalism between the world wars that depended on German farmers adopting what Bramwell called bio-dynamic agriculture in which they would eschew the use of artificial fertilizers. Although Steiner was not part of the Nazi party, some Nazi officials, such as Rudolph Hess, Hitler's deputy, and Walther Darré, Nazi Minister of Agriculture, thought Steiner's views could improve German agriculture. Despite their support, the Nazi government outlawed such ecological views by 1942 on the fear that those plans would turn Germans away from science and lead people in other nations to lower their opinions of Germany's culture.[45] The point was that an authoritarian government could jettison any environmental plan that conflicted with its needs, real or imagined.

Conclusion

Despite Tocqueville's arguments that the social conditions in America threatened democracy, the works of Emerson and Thoreau along with the efforts of the American missionaries to protect Native Americans illustrated the ways some Americans restrained those tendencies. In addition, there were major educational theorists in each general period of American educational history who offered ways schools might minimize those dangers. As noted in the introduction, these individuals include William Torrey Harris, John Dewey, and Gregory Bateson.

The ideas of these individuals fit the description Bramwell gave of the mindset of thinkers who worked during the turn of the twentieth century from the last years of the nineteenth century and favored an ecological perspective. She noted that these people shared three important qualities. First, they rejected dualisms or separations that implied people were independent of the environment in which they lived. Second, they embraced the notion that everyone and everything was part of the one earth. Third, they rejected the then contemporary political system or the then conventional patterns of beliefs even though they joined the mainstream of ideas occasionally.[46]

Although these philosophers did not hold to an ecological perspective, they offered increasingly wide definitions of democracy that offered ways to improve the social life of everyone and everything. For example, Harris offered a definition of democracy that showed how

DOI: 10.1057/9781137484215.0003

people were tied to the social conditions within which they lived. While the New England transcendentalists considered freedom as something a person sought apart from society, Harris argued that social progress made human freedoms possible. Dewey built on this notion by defining democracy as a way of life rather than a political system. For Dewey, democracy was a mode of conjoint communicated experience. Bateson went beyond Harris and Dewey to show that people had to realize that they were part of the environment and could not use nature to solve their problems. Both Harris and Dewey had asserted that people could shape the environment to fit their needs. Bateson considered such an approach to be immoral because it represented an unwillingness to recognize the connections that existed among things in the world.

The subsequent chapters will explain these points about Harris, Dewey, and Bateson and show how they offer a foundation for ecological sensitivity. The conclusion will return with some analysis of contemporary efforts to advance ecological sensitivity to show if or how Harris, Dewey, and Bateson provided a foundation for those perspectives.

Notes

1   David John Frank, Karen Jeong Robinson, and Jared Olesen, "Global Expansion of Environmental Education in Universities," *Comparative Education Review*, vol. 55, no. 4 (November 2011): 546–573, Stable URL: http://www.jstor.org/stable/10.1086/661253, accessed 4 October 2014.

2   Anna Bramwell, *Ecology in the 20th Century: A History* (New Haven: Yale University Press, 1989), 3–6, 39.

3   V. M. Spalding, "The Rise and Progress of Ecology," *Science*, vol. 17, no. 423 (6 February 1903): 201–210, Stable URL: http:www.jstor.org/stable/1629840, accessed 5 October 2014.

4   Stephen A. Forbes, "The Humanizing of Ecology," *Ecology*, vol. 3, no. 2 (April 1922): 89–92, Stable URL: http://www.jstor.org/stable/1929143, accessed 4 October 2014.

5   Richard S. Miller, "Observations on the Status of Ecology," *Ecology*, vol. 38, no. 2 (April 1957): 353–354, Stable URL: http://www.jstor.org/stable/1931696, accessed 4 October 2014.

6   Leo Marx, "American Institutions and Ecological Ideals," *Science*, vol. 170, no. 3961 (27 November 1970): 945–952, Stable URL: http://www.jstor.org/stable/17315541, accessed 4 October 2014.

7   Ibid.

DOI: 10.1057/9781137484215.0003

8   James Wheeler and Nobuo Shimahara, "Toward and Ecological Perspective in Education: Part 1," *The Phi Delta Kappan*, vol. 55, no. 6 (February 1974): 393–396, Stable URL: http://www.jstor.org/stable/20297592, accessed 4 October 2014.

9   Susan M. Leeson, "Philosophical Implications of the Ecological Crisis: The Authoritarian Challenge to Liberalism," *Polity*, vol. 111, no. 3 (Spring 1979): 303–318, Stable URL: http://www.jstor.org/stable/3234311, accessed 4 October 2014.

10   Tom Crawford, Editor's note to John Locke, *The Second Treatise of Government and a Letter Concerning Toleration* (1946 repr., Mineola, NY: Dover Publications, 2002), iii–iv.

11   John Locke, *The Second Treatise of Government and a Letter Concerning Toleration* (1946 repr., Mineola, NY: Dover Publications, 2002), 12–14.

12   Ibid., 14–18.

13   Peter S. Onuf, *Statehood and Union: A History of the Northwest Ordinance* (Bloomington: Indiana University Press, 1978), 1–20.

14   Daniel Feller, *The Public Lands in Jacksonian Politics* (Madison: University of Wisconsin Press, 1984), 14–38.

15   Angela R. Riley, "The History of Native American Lands and the Supreme Court," *Journal of the Supreme Court History*, vol. 38 (3 November 2013): 369–385, DOI: 10.1111/j.1540-5818.2013.12024.x. ISSN: 1059–4329, accessed 16 October 2014.

16   David R. Kimberly, "Cherokees and Congregationalists vs. Georgia and Andrew Jackson: The Attempt to Prevent the Trail of Tears," *International Congregational Journal*, vol. 11, no 1 (Summer 2012): 91–115.

17   Ibid.

18   Alexis de Tocqueville, *Democracy in America*, trans. Arthur Goldhammer (New York: Library of America, 2004), 371–391.

19   Ibid., 617–628.

20   Ibid., 581–586.

21   Ibid., 283–291.

22   For a statement concerning Tocqueville's continued relevance, see James R. Abbott, "Whither Tocqueville in American Sociology?" *The American Sociologist*, vol. 38, no. 1 (March 2007): 60–77, Springer, Stable URL: http://www.jstor.org/stable/27700490, accessed 29 March 2015.

23   John Stuart Mill, "M. de Tocqueville on Democracy in America," in *The Philosophy of John Stuart Mill: Ethical, Political, and Religious*, ed. Marshall Cohen (New York: Modern Library, 1961), 121–184.

24   Tocqueville, *Democracy in America*, 3–7.

25   Mill, "M. de Tocqueville on Democracy in America," 121–184.

26   Ibid.

27   J. Hector St. de Crèvecoeur, *Letters from an American Farmer* (1783 repr., Mineola, NY: Dover Books, 2005), 23–43.

DOI: 10.1057/9781137484215.0003

28   Tocqueville, *Democracy in America*, 21–29.

29   Ralph Waldo Emerson, "Nature," in *The Complete Essays and Other Writings of Ralph Waldo Emerson* (New York: The Modern Library, 1950), 3–44.

30   Henry David Thoreau, *Walden; or, Life in the Woods* (1854 repr., New York: Dover Publications, 1995), 205–206.

31   Bill McKibben, "What Would Thoreau Do?" *New York Review of Books*, vol. 61, no. 11 (19 June 2014): 50–52.

32   Donald Worster, *Nature's Economy: A History of Ecological Ideas*, 2nd edition (New York: Cambridge University Press, 1994), 67–70.

33   Ibid., 71–74.

34   Gifford Pinchot, "How Conservation Began in the United States," *Agricultural History*, vol. 11, no. 4 (October 1937): 255–265, Stable URL: http://www.jstor. org/stable/3739483, accessed 2 January 2015.

35   Ibid.

36   Ibid.

37   Ibid.

38   J. Leonard Bates, "Fulfilling American Democracy: The Conservation Movement, 1907–1921," *The Mississippi Valley Historical Review*, vol. 44, no. 1 (June 1957): 29–57, Stable URL: http://www.jstor.org/stable/1898667, accessed 2 January 2015.

39   William James, *Talks to Teachers on Psychology; and to Students on Some of Life's Ideals* (1899 repr., New York: W.W. Norton, 1958), 150–152. For further descriptions of the attitudes of the people who resisted conservation and lived in the forests and preserves, readers might also see Karl Jacoby, *Crimes against Nature: Squatters, Poachers, Thieves, and the Hidden History of American Conservation* (Berkeley: University of California Press, 2003).

40   Gordon G. Whitney, *From Coastal Wilderness to Fruited Plain: A History of Environmental Change in Temperate North America 1500 to the Present* (Cambridge, UK: Cambridge University Press, 1994), 227–249.

41   Hal S. Barron, *Those Who Stayed Behind: Rural Society in Nineteenth-Century New England* (Cambridge, UK: Cambridge University Press, 1984), 1–30.

42   Mark David Spence, *Dispossessing the Wilderness: Indian Removal and the Making of National Parks* (New York: Oxford University Press, 1999), 115–132.

43   Franz-Joseph Brüggemeier, Marc Cioc, and Thomas Zeller, "Introduction," *How Green Were the Nazis? Nature, Environment, and Nation* (Athens: Ohio University Press, 2005), 1–17.

44   Charles E. Closmann, "Legalizing a Volksgemienschaft," in *How Green Were the Nazis? Nature, Environment, and Nation*, ed. Franz-Joseph Brüggemeier, Marc Cioc, and Thomas Zeller (Athens: Ohio University Press, 2005), 18–42.

45   Bramwell, *Ecology in the 20th Century*, 195–208.

46   Ibid., 237–248.

DOI: 10.1057/9781137484215.0003

2
Developing Freedom within Social Institutions: William Torrey Harris and the St. Louis Hegelians

Abstract: *W. T. Harris and his colleagues in the St. Louis Movement wanted schools to show children how social restraints enhanced their freedoms. In this way, they created a definition of democracy that differed from that of the New England Transcendentalists. Although Harris believed people should use nature to enhance their well-being, he thought the recognition that everyone lives within a web of systems would prevent people from wasting those valuable resources. To offset individualism, materialism, and conformity, he turned psychological growth toward spiritual development, but he used philosophy and reason rather than religion to aid in this process.*

Keywords: consciousness; Hegelianism; self-activity

Watras, Joseph. *Philosophies of Environmental Education and Democracy: Harris, Dewey, and Bateson on Human Freedoms in Nature.* New York: Palgrave Macmillan, 2015.
DOI: 10.1057/9781137484215.0004.

A group of philosophers called the St. Louis Movement were able to extend conceptions of democracy in ways that could correct Tocqueville's fears that the American tendencies toward materialism, individualism, and conformity would threaten democracy. They made the corrections by turning to the ideas of the German philosopher Georg Wilhelm Friedrich Hegel. The most prominent member of the group was William Torrey Harris. Serving as a school administrator, Harris applied Hegel's ideas to suggest that schools could help students use the advantages and the limitations imposed by existing institutions to enhance their growth and their freedom. He believed such a curriculum would enhance social progress.

While Harris and the other St. Louis Hegelians were interested in nature study, they followed the then traditional view that people should use nature to enhance human welfare. Although they did not extend their thinking into something resembling an ecological movement, Harris has a place in a book about schools and ecological thinking because he tried to show that people lived within complex, interrelated systems and that school subjects had to reflect those relationships. In this way, Harris broke with the individualism that characterized the philosophy of New England Transcendentalists and set the stage for schools to show students that people were connected intimately to the environment in which they lived.[1]

Lawrence Cremin considered Harris to be the nation's first philosopher of education. According to Cremin, the reason that Harris towered above the other educational leaders of the nineteenth century was because he tried to explain the rational basis of the institution of the public school and to provide a theoretical basis for translating the mandate for universal schooling into practical reality during the decade after the U.S. Civil War. Although Harris believed that people grew when social institutions made the freedom of individuals their goal, Cremin criticized Harris for placing more emphasis on school room order than on student freedom. Cremin believed Harris's concern for organization inspired the reforms the progressive educational movement enacted after he died.[2]

Other commentators found that Harris made significant contributions to American social ideas. For example, Michael H. DeArmey and James Good claimed that the effort Harris made to apply the ideas of Hegel to social and educational problems enhanced the development of American thought. DeArmey and Good added that the St. Louis Hegelians maintained such philosophic rigor that they became important participants

DOI: 10.1057/9781137484215.0004

in the philosophical and educational debates of the late nineteenth century.[3]

Using Hegel to understand the relation of all things on earth

According to Denton J. Snider, one of the founding members of the St. Louis Movement, the St. Louis Movement improved philosophic thought in the United States because they took the ideas of Hegel as their spiritual guide. From Hegel, they gained the view that the Idea or the Spirit was the source of all things on earth. This made them look beyond specific events to find some higher influence that explained why things happened as they did. In addition, Denton praised Harris for showing that education was the important avenue of social reform.[4]

Harris illustrated the two tendencies that Snider found important when he gave a speech to the National Education Association (NEA) in 1870. Two years earlier, Harris had assumed the position of superintendent of the St. Louis public schools. In this speech, Harris demonstrated how philosophy could illuminate the relation of classroom practices to national events, and he showed how classroom practice could strengthen the spirit of democracy. The subject of the speech was a defense of the use of textbooks in schools. At the time, a popular classroom technique was the object method of teaching that spread from the Oswego, New York, normal school. In this method, teachers asked the students to observe various objects, remember their characteristics, and answer questions about them.

To begin his address, Harris claimed that the U.S. Civil War was part of the process of evolution that resulted from the natural dialectic between nations. He asserted that the Union won because Northerners held the view that industrial progress would further humanitarian goals while Southerners lost because they favored aristocratic models of serfdom. Harris added that the war forced the Southerners to accept industry to advance their society. From this observation, Harris argued that history changed as a result of ideas that were moving toward the enhancement of people's freedom; the skills of the leaders were unimportant. In this speech, Harris used the war as a metaphor to illustrate which classroom practices were appropriate for a democracy. Although teachers considered lectures and demonstrations as the best means to transmit

DOI: 10.1057/9781137484215.0004

knowledge, Harris predicted that educators would see that textbooks provided the means to advance human freedoms. Textbooks enabled the children to acquire information on their own from a trustworthy source. He added that when students depended on teachers to learn skills and information, they came to think that as adults in society they should depend on human leaders.[5]

Although Harris's speech to the NEA suggests that he made philosophy a central aspect in school reform, he did not move to St. Louis in 1857 to expand his involvement with education and philosophy. His aim was to make his fortune in farming. He quit Yale University after two years of study; however, he could not succeed in farming or in building a business in St. Louis. Accordingly, he turned to teaching elementary school in 1858 at the age of 22. His rise was rapid. The next year, he became the principal of Clay School, and the St. Louis school board appointed him superintendent of St. Louis schools ten years later, in 1868.[6]

The position of superintendent was a rare opportunity for Harris. In 1870, there were only 29 city superintendents in the United States even though there were 226 cities with populations exceeding 8,000 people. More important, superintendents in city school districts had many opportunities to bring about reform that most rural districts could not afford. Most rural districts had one-room schoolhouses; however, superintendents in cities could establish graded schools, high schools, and a variety of educational innovations such as kindergartens. In addition, a superintendent in a city school district could implement business like administrative techniques. Accordingly, Harris had opportunities to determine whether or how these innovations would advance democracy.[7]

The idea that educational innovations had social effects came to Harris from Henry Conrad Brokmeyer, who introduced him and his friends in the St. Louis Movement to the ideas of Hegel in 1858. Following Brokmeyer's teaching, Harris and his friends used Hegel's ideas to answer the important questions of the day. Philosophy became for them the most practical of all types of knowledge. As noted earlier, they connected philosophy to school teaching and to school management.[8]

Another advantage that St. Louis offered Harris was that he and his colleagues in the St. Louis Movement believed they enjoyed unique opportunities to grow intellectually in the city. They agreed that their knowledge of Hegel's ideas would help them contribute to the world. According to the U.S. Census, St. Louis was then the eighth largest city

DOI: 10.1057/9781137484215.0004

in the United States with a population of a little more than 160,000 residents in the 1860s. Believing the rapid growth of their city would make it a great city of the world, Harris and his colleagues wrote books on education, literature, religion, philosophy, economics, and art. The books came from discussions the group had organized, and they illustrated the faith Harris and his colleagues had that their philosophical inquiries could help solve important problems facing a growing industrial world.[9]

For his part, Harris made three important contributions to education and to the then contemporary intellectual life. First, he took advantage of philosophy to make school lessons relevant to social changes. Second, he built on a conception of psychology that blended physical and spiritual growth. Third, he devised a curriculum that organized subject matters to reveal to students the places people occupied in the institutions that made up society. Through these efforts, Harris helped people recognize that their patterns of thoughts influenced their ethical conceptions. These contributions offered an alternative to the dangerous tendencies of individualism, materialism, and conformity that Tocqueville feared threatened American democracy. The following sections will explain how Harris's philosophical innovations could offset the dangerous tendencies of democracy.

Reducing individualism through a curriculum related to social changes

Harris reduced the tendencies toward individualism in two ways. First, he adapted the curriculum to the social changes that created an integrated industrial society in the United States after the U.S. Civil War. Second, he advocated a model of psychology that built on the notion of human growth through self-activity that showed the children their development was tied to their movement toward spiritual enlightenment. In this effort, the children needed the guidance of trained teachers who were aware of the implications in practical lessons.

The first way Harris reduced the dangers of individualism was to adapt the school curriculum to the period of momentous change that began in the United States with the end of the U.S. Civil War. The popular view before the war had been that personal individualism was the best expression of freedom and the autonomy of small towns made it possible. During the 1870s and 1880s, the growth of businesses and the spread of

DOI: 10.1057/9781137484215.0004

railroads connected citizens who lived far apart, and officials introduced managerial forms of administration to cope with the expansion.[10]

Intellectual shifts accompanied the social and political changes. Before the U.S. Civil War, the popular conception of philosophy was that individuals should avoid social constraints. This idea appeared in the idea of local control within small communities. The New England Transcendentalists extended this notion with the view that each American should discover his or her divinity apart from social connections. Harris broke with the Transcendentalists and took a view consistent with the increased social control that came with the end of the U.S. Civil War. This control came from several sources such as the increased authority of the federal government, the rise of industrialism, and the growth of cities. Harris turned what the Transcendentalists saw as a disadvantage into a benefit. He argued that human beings made their potential real through social institutions. Although the Transcendentalists had argued that freedom appeared when people escaped the impositions of society, Harris and the Hegelians countered that a person was free when he or she rose above the social context and saw how mutual impositions within society made freedom possible.[11]

In fairness, the members of the St. Louis Movement and the Transcendentalists influenced each other through their disagreements. These groups attended meetings together where they exchanged ideas. Harris knew and admired Ralph Waldo Emerson and Bronson Alcott. Harris invited them to address the Philosophical Society in St. Louis, and he was an important figure at the Concord School of Philosophy from 1879 to 1888.[12]

Although Harris did not want students to follow their impulses freely as Transcendentalists may have wished, he avoided authoritarian efforts to force teachers or students to serve the needs of the organization. Instead, he tried to shape the school practices in ways that served the teachers and the students within the limits of the school's fundamental social mission. This fit Harris's belief that people acted ethically when they understood how institutions provided for their greater freedom, and leaders acted ethically when they organized institutions to direct people to enlightenment. For example, in St. Louis, Harris began keeping records of the students' attendance and their tardiness. The initial reason for the record keeping was that the state of Missouri appropriated money to schools for children aged between five and twenty years old, and the information could justify state support. Nonetheless, Harris wanted the

DOI: 10.1057/9781137484215.0004

records to inspire the teachers and the parents to speak to the children about the importance of education. This offered parents and students opportunities to consider the value of regular school attendance.[13]

As a school superintendent, Harris's involvement in political affairs suited his view that philosophy was a practical activity. He served as president of the National Education Association, and U.S. President Benjamin Harrison appointed Harris in 1889 to be the fourth U.S. Commissioner of Education.[14]

When Harris became the U.S. Commissioner of Education, he adopted a new definition of democracy. As Commissioner, Harris collected statistics and evidence to show innovations that could improve education. As an important figure in the NEA, he expected it to spread the information to school people across the nation, and he believed that reasonable people would adopt those methods deemed to be best for everyone. This definition of democracy allowed individual educators to act freely, but the institutions of government offered information and training that could help those educators improve their own schools.

One example is illustrative. In 1893, Harris was a member of the NEA's Committee of Ten, which recommended the curriculum for high schools. It consisted of the subject areas Harris deemed most important: classical languages, Latin and Greek; modern languages; mathematics; physical science; natural history; history; and geography. Three years later, Harris served as chairperson of the NEA's Committee of Fifteen, which outlined a curriculum for elementary schools that paired well with the high school program. Although there was no requirement that school officials follow these suggestions, Harris was able to announce by 1904 as U.S. Commissioner that high schools in the United States followed the curriculum plans outlined by the Committee of Ten.[15]

Reducing the tendencies toward individualism by directing psychological growth toward spiritual development

The second way Harris reduced the problems of individualism was through a model of psychology that built on the notion of human growth through self-activity. Although the term implied personal independence, it meant that the children's development was tied to their movement toward spiritual enlightenment. Since Harris believed this process could

DOI: 10.1057/9781137484215.0004

be described, he argued that it was possible to train teachers to guide the students to what Harris considered the higher levels of thinking. This meant the teachers had to be familiar with the psychological development of their students and with the ways the arrangements of social institutions facilitated this growth. Although the term, "self-activity," implied the organism was independent, this process was distinct from individualism.

In the last quarter of the nineteenth century, there were two popular views of psychology. One was the older model of faculty psychology, and the other was a newer view of physiological psychology. The older view described faculties as aspects of mental life that students could exercise with difficult puzzles in the ways that athletes strengthened specific groups of muscles with appropriate training. The newer psychology made thinking a habitual response to a stimulus. Self-activity offered a third option.[16]

Although Harris thought that all living creatures engaged in self-activity, he added that this instinct changed as it moved to higher levels. For example, a plant exhibited self-activity when it drew sustenance from soil, water, and light in order to grow. Infants engaged in self-activity when they observed something and felt an emotional response. As people matured, they received the results of the perceptions that the members of their society had accumulated. These impressions lifted them above the then present moment and afforded them directions for self-control and ethical behavior. Harris gave the name "culture" to the accumulated reflections of the members of the society. Since human beings could avail themselves of a culture, it distinguished them from animals because it set them on a path to an ideal that existed beyond their experiences.[17]

According to Harris, the concept of self-activity implied that human beings were born as animals but became spiritual beings. This evolutionary process began with infants noticing things around them. Once they focused attention on something, they gathered information about it and excluded impressions from other objects. The process of analysis began when the children isolated things they had observed about the object. This step informed the children about the object and led to the step of synthesis, in which the children saw connections between the object and other things in the universe. The final step was a form of philosophic knowing or insight in which the person could see the ways the thing participated in God's final purpose in the universe.[18]

DOI: 10.1057/9781137484215.0004

Although the steps listed here applied to concrete objects, Harris followed those steps when he described the ways that self-activity directed abstract, academic lessons. In both cases, the steps appeared to focus attention, gather information, analyze the aspects of the object, and make connections to other things. For example, in a lesson on grammar, the mental operations began with introspection. First, the children had to recall their past experiences which involved the object to which a word applied. This meant that any lesson about any word began with recalling information about that specific object and separating it from other information about any other object. Second, the children had to recognize the way the word functioned in a sentence. This was the process of synthesis because the children saw the relation of the word to other words. These impressions came from the processes of attention and analysis listed in the earlier paragraph. They illustrated self-activity because the children did these actions for themselves.[19]

The important point in Harris's description of the grammar lesson is the way the teachers' understanding of psychology helped them guide students to move from lower levels of comprehension to higher levels. Each step differed from the others. While recalling the relation between an object and a word depended on introspection, seeing the function of a word in a sentence required the children to think logically. This meant to Harris that teachers had to realize that the higher steps of cognition included the lower steps, but those steps opposed each other as well. The reason teachers had to know these steps was that they might try to train those abilities to work in harmony with each other. Such an effort would fail because children could not use the same style of thinking in recalling an object, matching it to a word, and recognizing how the word functioned in a sentence. Each step depended on a different style of thought and the children could not understand how the lower activities blended together until they reached the higher levels and made a synthesis by acquiring the skill of reading. Harris feared that untrained teachers would force the children to repeat drills relating experiences to words or words to other words in a sentence. Such mindless repetitions would arrest the children's development.[20]

Following a similar idea, Harris thought that each school subject required its own manner of thought. Accordingly, he opposed efforts to mix subjects indiscriminately because this would interrupt the students' efforts of mastering the pattern of thinking appropriate to any subject area. For example, Harris thought the subject of grammar required logical thinking; however, the study of literature opened aesthetic experiences wherein the student vicariously experienced feelings grow into deeds that usually involved a clash with some institution and called for a tragic or comic solution. Harris did not want teachers to pause in a discussion of a poem, play, or story to illustrate a grammatical point in a specific sentence. Such a disturbance would harm the aesthetic experience the students could derive from the literature and interfere with the logical patterns the grammar introduced.[21]

Arranging the curriculum into a series of activities that merged into syntheses, Harris conceived of the curriculum as opening the students to spiritual awakenings. Although the school was not to undertake the work of the Church, he believed it could prepare the way for religion by illuminating the roles of the other social institutions. In order for the school to have this effect, each institution had to remain separate. If the children did not acquire fully the way of thinking from each institution, the blending would be a mixture not a synthesis.

Harris applied this rationale to resist the inclusion of a newly popular model called "manual training." Advocated by Calvin Woodward, manual training was an effort to teach students how to work with tools and to make practical judgments. Woodward wanted it to augment academic studies rather than become a vocational trade program. Lessons might include identifying types of files and using them in graded steps to make a piece of iron square and to turn round holes into square ones.[22]

Harris disagreed with Woodward's proposals because the skills the students learned in manual training came from sense perceptions. Although such impressions were important, Harris did not think they were the proper realm of school studies. For Harris, the aim of academic subject matter was to show the general principles that underlay things in the world. To do this, the students had to learn to unite heterogeneous things, such as leaves, acorns, and branches, into patterns that enabled them to recognize oak trees. This required a type of thinking that moved from concrete objects to abstract concepts. For example, Harris noted that mathematics began with material objects, but the idea of quantity offered ways to treat objects that did not exist. Accordingly, arithmetic might assist a trade, such as carpentry, but it called for understandings of relations that were not concrete but could have concrete applications, such as theorems in geometry. The same was true of learning to read or studying literature. These efforts went far beyond doing something with tools because they moved into the realm of human ideals. These

DOI: 10.1057/9781137484215.0004

thoughts had practical application because they indicated the possible solutions for human problems. Nonetheless, Harris was willing to accept drawing as a school subject even though it was often a part of manual training. The reason was that drawing could go beyond the facile use of pencil and paper and offer students the opportunity to recognize beautiful and graceful forms. Such aesthetic sensibilities served all people not just carpenters or metal workers.[23]

For Harris, the justification for separating religion, manual training, and academics was that the institutions of church, work shop, and school employed different methods and sought distinct ends. When Harris explained the different roles of social institutions, he began by comparing schools and families. He claimed that the special work of the school was to teach students letters and civil manners. Although families taught manners, they did it by having the child share in the lives of the other family members. On the other hand, schools helped students develop independence. They learned to relate to other independent individuals and to people with authority over them. In schools, students worked at prescribed tasks and they practiced virtues of regularity and punctuality, which made obedience an essential part of the school as an institution.[24]

It appeared that Harris contradicted his own philosophy when he established the first public school kindergarten in the United States in 1873. This was not the case. As superintendent of St. Louis schools, Harris resisted the kindergarten movement because he believed those teachers used children's play for serious ends. He thought these teachers distorted the nature of play and the free exercise of caprice that he believed the children required to develop their personalities. To his surprise, Harris discovered that kindergartens differed in that they furnished an ingenious graded course of exercises that helped the child take interest in serious activities. On these grounds, Harris came to see the kindergarten as a means of transition from the family to the school.[25]

According to Harris, kindergartens offered two levels of activities. The lower level built on several gifts or objects that Froebel and other teachers had devised for the children. Froebel began with a woolen ball, a cube, a cylinder, and sphere. These gifts provided the basis on which kindergarten teachers made 20 or more gifts the children could use to make designs of things such as houses and animals. The kindergarten expressed this in a slogan: all things appear in all things. It was a view based on a theory of crystals as the building blocks of matter. The aim

DOI: 10.1057/9781137484215.0004

was to inspire children to recognize how all parts of nature and society fit together.[26]

The higher level was where Harris thought the kindergarten fulfilled its potential. This was the area of plays and games. With the gifts, the children exerted their power over material objects. In the plays and the songs, the children became aware of their social selves because they imitated the productive activities in society and recognized their relationships to other human beings.[27]

Although Harris accepted the kindergarten into schools, he remained firm in his opposition to include religion in the curriculum. Harris claimed such a merger was ill advised. He noted that schools taught children about conventional behavior, offered instruction in communication, and suggested an intellectual view of the world. Although the aims of the school subjects of literature and science conformed to the expectations of religion, the ways of thinking in the subject matters differed from the patterns expected in religion. In academic matter, students learned to recognize how human beings used nature to their advantage. They did not perceive the authority of the Divine in this process. In addition, religion conveyed its messages through the acceptance of the authority of the Bible while the school encouraged the students to think critically and examine all the evidence that was available. Placing religion in the school might encourage students to treat it as critically as they would a history lesson or scientific experiment.[28]

The point is that Harris devised explanations of the school curriculum that blunted the popular tendencies toward individualism by defining individual freedom as the ability to recognize the distinct aspects of human culture and to understand how they fit together because of their unique aspects. In this way, the curriculum set the foundation for the recognition of the ways people fit within their environments. This understanding could become the basis of an ethical framework favoring ecological protection.

Avoiding the tendency to conformity with a curriculum that revealed people's places in society

Despite Harris's opposition to integrating home care, religion, and vocation in the school curriculum, he believed schools should help students learn to function in the institutions that made up society as they traveled

DOI: 10.1057/9781137484215.0004

toward a spiritual awakening. He arranged those institutions in hierarchal order beginning with families, moving to civil society, adding the state, and concluding with the Church. The school appeared within the other parts of civil society. According to Dewey, the innovation that Hegel offered was a response to the exaggerated complaints of Jean-Jacques Rousseau about the tendencies of civilization to enslave humankind. Dewey noted that Hegel had advanced the view that social institutions brought children to the present level of civilization. Accordingly, Dewey praised Harris for following Hegel's views to construct the curriculum around questions of human development. In doing this, Harris dropped the idea of curriculum as the logical presentation of subject matter. He replaced it with a conception of curriculum as a progression of experiences that led the child to appreciate the features of his or her society.[29]

Harris believed that limiting the curriculum to five subjects would best illuminate the features of society. He called these subjects the windows of the soul because they showed how human beings used nature for their benefit, and they opened on to what Harris believed were the divisions of life in society. Two of the subjects described aspects of nature. One was inorganic nature, which included arithmetic and natural philosophy to survey whatever involved time or quantity. The other was organic nature, which included geography and natural history that showed how human beings used nature to provide food, clothing, and shelter. Three of the subjects related to human society. The first was grammar, which showed logical organization; the second was history, which illuminated the willpower of the nation; and the third was aesthetic, which included literature and singing. Although Harris wanted all five subjects to remain in symmetrical arrangement throughout the students' progress from elementary school through college, he thought they should appear in differing amounts and in increasing complexity.[30]

The subjects became more advanced as the students moved through the grades. For example, in secondary school, arithmetic became algebra, geometry, and trigonometry while language study became Latin, Greek, and French or German. In college, algebra became analytical geometry and differential calculus while language study continued work on the classical languages and moved on to studies of Plato and Aristotle.[31]

Since the curriculum was to teach students about their function in society, the various courses showed them the intricate relationships among thoughts and actions. For example, Harris contended that there were two extremes in the course of study. Mathematics dealt with things

DOI: 10.1057/9781137484215.0004

in a mechanical aspect while literature looked at the human aspects of life. In learning arithmetic, the students had to distinguish between quality and quantity. For example, the children could count three oxen by recognizing they shared the same quality. If there were two oxen and one horse, the children had to separate the items by their distinct qualities. This meant that the children had to learn two different methods of thinking with the idea of quality being more advanced than that of quantity. On the other hand, literature revealed the desires of human beings. It gave students opportunities to think about the ways sin and crime upset the divine order of the world. In these efforts, the student came to know the inner workings of other human beings.[32]

In making these descriptions of the functions the subjects served, Harris blunted the tendencies toward conformity that existed in society. Instead of moving things toward similar goals, he carefully distinguished how differences advanced thought and understanding. Such complexity should enable students to pause before reacting emotionally to situations and ignoring larger ethical considerations.

Avoiding the tendencies toward materialism by learning about the social value of private property

As the students learned about contradictory ways of thinking, Harris hoped that they would come to recognize how the mutual impositions in life made freedom possible. Harris's view of self-activity contradicted the notion that sensations directed people's desires. For example, while an empirical psychologist might contend that a hungry person saw a piece of fruit and ate it because the sensation of taste and the need for food directed his or her action, Harris took another view. He claimed that the taste of the fruit or the sated pangs of hunger did not exist before the person ate the fruit. Thoughts of these sensations arose because the person's mind made abstract leaps associating various actions with results. The important point for Harris was that mental laws did not determine this sequence. The person was free to impose into this sequence a moral quality such as the concept of property. In this case, the person could ask who owned the fruit, and the person could recall that society would not exist if people violated each other's rights to property. Once a person raised these moral questions, he or she would recognize which actions were most beneficial to society. From this type of moral thinking, Harris

DOI: 10.1057/9781137484215.0004

believed people could choose to act on motives that would reinforce the common good and resist the drive to accumulate material wealth that seemed to permeate American society.[33]

As this example shows, Harris considered self-activity as an essential characteristic for human freedom. In this regard, Harris paraphrased Plato to argue that people moved through self-activity toward a perfect reflection of the Creator who embodied love and existed above time and space.[34]

Although self-activity moved toward the Creator, Harris did not contend that this movement was linear or direct. According to Harris, self-activity moved through three levels of thought. Once again, Harris followed Plato's ideas and described the lowest level as the stage when a person considered objects as independent of each other. At the second stage, the person came to understand that the world depended on the mutual relation of things. In this stage of thought, people looked to the forces that held things together, such as gravity, and these became more important than objects that existed independently. At the highest level of thinking, the person discovered total systems and self-determining principles that shaped independent beings. Harris placed the discovery of self-activity as the highest form of thought because it implied that a self-active being came to recognize that it was following the form of the absolute.[35]

For Harris, the mutual impositions involved in the concept of private property built upon yet modified the pursuit of material gain that threatened democracy. On the one hand, the concept of private property seemed to restrict wealth to a few people. On the other hand, private property was a necessary condition for social progress.

Harris expressed his views on private property in response to Henry George's publication in 1879 of *Progress and Poverty*. In that book, George contended that private property restricted natural resources because a few people had taken possession of the available land. George argued that these conditions allowed for the evils of unjust and unequal distribution of wealth. Since a few people owned land, the rest of the population lived as tenants. Unfortunately, both groups refused to cooperate to improve life for everyone. The solution George proposed was a single tax on land values that would ensure the land was turned to some use that was profitable for everyone.[36]

In one sense, George followed the model that Harris had set; he sought the one factor to which he could reduce the many problems he described. In another sense, George differed from Harris's approach because he did not find the cause in the movement of an abstract essence toward its fulfillment. George was satisfied that the arrangements of land ownership was the physical catalyst promoting the spread of poverty.

Writing an essay to criticize George's book, Harris argued that private property was essential for social progress. Following his own model of thinking, Harris claimed that private property was an essential aspect of spiritual life because it increased mutual interdependence among people. This interdependence developed through the inventions of railroads for transportation or factories for material production. Although Harris acknowledged that some people had garnered enormous fortunes from these innovations, he felt that these industries offered more benefits to humankind than harms from the inequality of income. In fact, it appeared that the gains those people made were minuscule compared to the wealth they created for the rest of society, and the successful innovators had bested several competitors in those fields who received nothing for their efforts. Had society rewarded the competitors equally, the costs would have exceeded what the entrepreneurs earned on their own. Thus, Harris was content to reward individual successes provided the wider society benefitted from that success more than the individual did. Under these conditions people could enhance freedom for each other.[37]

In making his statements about property, Harris seemed to believe that the pattern of mutual impositions would limit what owners could do with their property. Because a person's freedom to own property enabled him or her to fulfill their obligations to society, they could not use their ownership to justify irresponsible actions. For example, a person needed private property to produce goods to share with other people. Without some possessions, a person could not participate in the labor of the race or receive anything in return. This did not mean they could destroy those goods or accumulate more than was reasonable. For Harris, waste and greed injured the individual and the society by diverting property from its appropriate role in advancing human freedom.[38]

Harris made his observations about waste in his commentary on Dante's *Divine Comedy*. Acknowledging that few people found spiritual meanings in poetry, Harris excused the poets that could ruin their poems by introducing philosophy into them. Since Dante's poem was a religious poem, he could introduce significant spiritual lessons especially while he revealed the inner thoughts and feelings of sinners. As noted earlier, this was an important role of literature in school studies.[39]

DOI: 10.1057/9781137484215.0004

Critics disagreed with Harris's defense of existing social inequities. For example, when Merle Curti wrote the tenth volume of the Report of the American Historical Association's Commission on the Social Studies in 1935, he argued that Harris propounded a form of Hegelian philosophy that endowed men and women with noble destinies while it justified the social order in ways that subordinated those individuals to existing institutions.[40]

Despite Curti's criticisms, Harris offered a way to think about human psychology, school lessons, and society that showed the essential relationships existing among their parts. Not only did Harris move curriculum from a logical arrangement of subject matter, he constructed it in ways that showed how people advanced their freedoms by recognizing necessity of the restraints social institutions required. This observation could provide a foundation for an ethical framework that would restrict environmental destruction.

Conclusion

By the time Curti wrote his essay, Harris's ideas had fallen from prominence. Curti's criticisms were not the important element causing people to turn away from Harris's ideas. The reason Harris became outdated was that the study of philosophy and psychology changed in the twentieth century. According to George Santayana, the views of Harris and the members of the St. Louis Movement were the dominant view in America in the nineteenth century. Santayana attributed this dominance to the fact that American philosophy had been rooted in religion, which allowed Transcendentalists, Unitarians, and pragmatists to function within an atmosphere of German idealism. Under this model, consciousness was the faculty of mind that called upon memory to give meaning to events. Santayana argued that the critical attitude toward consciousness was the most influential element in this philosophical tradition. By the end of World War I, this critical attitude had grown to the point where philosophers joined psychologists in questioning whether consciousness existed. Santayana claimed that his teacher, William James, turned against the idea of consciousness on the grounds that no one could experience it. A person could know something about specific experiences, but consciousness was not one of those experiences. Santayana acknowledged that James's view seemed democratic because it removed

DOI: 10.1057/9781137484215.0004

the elitist notion that some people recalled things other people ignored; however, Santayana argued that the view that consciousness did not exist called the idea of truth into question. Nonetheless, philosophers went ahead and tried to perfect this pragmatic view of knowing.[41]

When philosophers and psychologists argued that consciousness did not exist, they made Harris and the St. Louis Hegelians appear to be out of date. After all, Harris and the St. Louis Hegelians had devoted their attention since the end of the U.S. Civil War to explanations of how consciousness developed. This change in philosophic ideas in America was the basis of the criticisms that greeted the text Harris published in 1899, *Psychologic Foundations of Education*. In the book, Harris tried to show the important educational factors in civilization and in schools that enhanced the evolution of the higher psychological faculties.

In reviewing *Psychologic Foundations of Education*, John Dewey acknowledged that Harris offered a helpful view of culture and of the ways people could enlarge their spirits. Dewey complimented Harris for offering insights into the underlying forces which influenced the aims of education. Despite this praise, Dewey complained that Harris's view of psychology ignored the renaissance then going on in psychology. When Harris wrote about human growth, he described it as a process of self-activity, which built upon an idea of a person functioning within a systematic, universal whole defined as reason. Dewey added that the newer psychologists, of whom he was one, discounted the spiritual factors that Harris valued. Instead, the newer psychologists considered growth to be a product of physiological factors. Nonetheless, Dewey agreed with Harris's view that individuals developed their latent, higher capacities within community life.[42]

For more than one half of a century, Harris dominated educational thought, and he offered alternatives to the notion of democracy that countered individualism, mindless conformity, and excessive material gain. Tocqueville had noticed that these tendencies derived from the circumstances of American democracy, and he feared they threatened democracy itself. Since Harris built his notion of education and human freedom on Hegel, he was able to suggest alternatives that limited those dangers. For example, Harris did not disdain individualism or advance conformity against freedom. Instead, he argued that freedom developed in the cradle of social restraints. For example, textbooks might limit information to things authors expressed, but the books enabled students

DOI: 10.1057/9781137484215.0004

to learn independently and to develop their own thoughts on the basis of what they had read. Students who learned entirely from teachers' lectures might hear misinformation and they could not reconsider those lessons in the quiet of their study or in the company of their fellows. In the same way, he argued that material gain was necessary but dangerous. By placing greed in an ambiguous position, he did not have to fight against it. He tried to turn it toward human good. In these efforts, Harris wanted students to learn about the many contradictions in social life. For this reason, he promoted reading literature that portrayed tragedy and vice. He wanted different institutions to remain separate. People thought differently while reading a newspaper, participating in a political campaign, or praying in church. Harris sought to preserve these differences in social institutions, and he wanted schools to help students recognize the arrangements found in the wider society. In these ways, Harris expressed his faith that people could see the truth if they honestly faced social circumstances. With this faith, democracy could save itself from its own destructive tendencies and thereby provide a basis for the preservation of the environment.

Notes

1   Readers should see John R. Shook and James A. Good, *John Dewey's Philosophy of Spirit* (New York: Fordham University Press, 2010), especially page vii.

2   Lawrence A. Cremin, *The Transformation of the School: Progressivism in American Education* (New York: Alfred A. Knopf, 1967), 14–20.

3   Michael H. DeArmey and James A. Good, "Introduction," in *Origins, the Dialectic, and the Critique of Materialism*, eds Michael H. DeArmey and James A. Good (Bristol, England: Thoemmes Press, 2001), vii–xx.

4   Denton J. Snider, *The St. Louis Movement in Philosophy, Literature, Education, Psychology with Chapters of Autobiography* (St. Louis: Sigma Publishing Co., 1920), 19–29.

5   William T. Harris, *The Theory of Education* (Syracuse, NY: C.W. Bardeen 1893), 12–13.

6   Kurt F. Leidecker, *Yankee Teacher: The Life of William Torrey Harris* (New York: Philosophic Library, 1946), 69–72, 153–166, 178, 245.

7   Elwood P. Cubberley, *Public School Administration: A Statement of the Fundamental Principles Underlying the Organization and Administration of Public Education* (1916 rev., Boston: Houghton Mifflin Co., 1929), 75–77.

DOI: 10.1057/9781137484215.0004

8   William T. Harris, *Hegel's Logic: A Book on the Categories of the Mind*. (1890, repr., New York: Kraus Reprint Co. 1970), xiii.

9   Henry A. Pochmann, *New England Transcendentalism and St. Louis Hegelianism* (Philadelphia: Carl Schurz Memorial Foundation, 1948), 19–21.

10   Robert H. Wiebe, *The Search for Order, 1877–1920* (New York: Hill and Wang, 1967), xiii–xiv.

11   Michael H. DeArmey and James A. Good, *Origins, the Dialectic, and the Critique of Materialism* (Bristol, England: Thoemmes Press, 2001), 14.

12   Pochmann, *New England Transcendentalism and St. Louis Hegelianism*; Shook and Good, *John Dewey's Philosophy of Spirit*, vii.

13   Leidecker, *Yankee Teacher*, 153–166.

14   Ibid., 456–459.

15   Henry J. Perkinson, *The Imperfect Panacea: American Faith in Education, 1865–1990* third edition (New York: McGraw-Hill, Inc., 1991), 136–137.

16   W. T. Harris, *Psychologic Foundations of Education: An Attempt to Show the Genesis of the Higher Faculties of the Mind* (New York: D. Appleton and Co., 1899), v–x.

17   Ibid., 228–239.

18   Ibid., 240–250.

19   Ibid., 325.

20   Ibid., v–x.

21   Ibid., 327.

22   Calvin Woodward, *The Manual Training School* (1897, repr., New York: Arno Press, 1969), 272–282.

23   William Torrey Harris, *Psychology of Manual Training*: A Paper Read before the Department of Superintendence, National Education Association, Washington, DC, March 7, 1889, https://archive.org/details/psychologymanuaooharrgoog, accessed 17 July 2014.

24   William T. Harris, "Vocation versus Culture; or the Two Aspects of Education," *Education: Devoted to the Science, Art, Philosophy, and Literature of Education*, vol. 13, no. 4 (December 1891): 193–206.

25   Ibid., 201–203.

26   Norman Brosterman, *Inventing Kindergarten* (New York: H.N. Abrams, 1997).

27   Harris, *Psychologic Foundations of Education*, 315–319.

28   William T. Harris, "Morality in the Schools" (Boston: Christian Register Association, 1889): http://archive.org/details/morality, accessed 13 July 2014.

29   John Dewey, "Culture Epoch Theory," in *A Cyclopedia of Education*, ed. Paul Munroe (New York: Macmillan Co, 1911), 240–242.

30   Harris, *Psychologic Foundations of Education*, 322–340.

31   Ibid., 340.

32   Ibid., 341–345.

33   Ibid., 120–127.

DOI: 10.1057/9781137484215.0004

34 Ibid., 36–37, 118.

35 Ibid., 32–37.

36 Henry George, *Progress and Poverty: An Inquiry into the Cause of Industrial Depressions and of Increase of Want with Increase of Wealth... The Remedy* (1879, repr., New York: Robert Schlkenbach Foundation, 1958), 3–16, 328–332, 403–408, 456.

37 W. T. Harris, *The Right of Property and the Ownership of Land* (Boston: Cupples, Hurd, & Co., 1887), 22–40.

38 W. T. Harris, *The Spiritual Sense of Dante's Divina Commedia* (Boston: Houghton, Mifflin, and Co., 1899), 68–69.

39 Ibid., 2–3.

40 Merle Curti, *The Social Ideas of American Educators*, with a New Chapter on the Last Twenty-Five Years (1935, repr., Totowa, NJ: Littlefield, Adams & Co., 1978), 310–347.

41 George Santayana, "Philosophical Opinion in America," *Third Annual Philosophical Lecture, Henriette Hertz Trust* (London: Oxford University Press, 1918).

42 John Dewey, "Harris's Psychologic Foundations of Education," *Educational Review*, vol. 16 (June 1898): 1–14.

DOI: 10.1057/9781137484215.0004

3
Pragmatism and Ecological Conservation: The Ideas of John Dewey

Abstract: *Dewey rose to prominence when Theodore Roosevelt started the conservation movement. Accordingly, Dewey criticized business corporations that destroyed virgin lands, polluted rivers, and wasted valuable resources seeking immediate profit. Tying these criticisms to the ways people should think, Dewey contended that people would reduce the dangers of individualism, materialism, and conformity if they sought satisfaction within activities they wanted to pursue. Further, he wanted schools to teach students to think in ways that benefitted the students and the society.*

Keywords: aesthetic sensibilities; experience; scientific method

Watras, Joseph. *Philosophies of Environmental Education and Democracy: Harris, Dewey, and Bateson on Human Freedoms in Nature.* New York: Palgrave Macmillan, 2015.
DOI: 10.1057/9781137484215.0005.

DOI: 10.1057/9781137484215.0005

During the last years of the nineteenth century, W. T. Harris and his colleagues expanded the notion of democracy. While the New England Transcendentalists had argued that people found their freedom and their divinity apart from society, Harris saw social restraints as essential conditions enabling people to develop their freedoms and achieve fuller realizations of their spirituality. Harris's view represented an expansion because it required a fulfillment of social customs that Emerson and Thoreau sought to escape. John Dewey turned Harris's notion of social restraints into a conception of democracy as a mode of associated living. In this way, Dewey was more explicit in showing how people improved their abilities and their accomplishments when they cooperated in activities of mutual interest.[1]

By expanding their notions of democracy, these philosophers reduced the dangers inherent in American values that Tocqueville believed threatened the benefits of their social system. In addition, they made possible an ethical framework that would make people sensitive to environmental preservation.

In fairness, the differences among the Transcendentalists, Harris, and Dewey were minor. For example, in the 1830s, Tocqueville noted that Americans constantly joined together to create hospitals, churches, and schools. From these activities, they learned to set large numbers of people to work toward a shared goal.[2] The benefits that society showered on individuals were not lost on Ralph Waldo Emerson. In his essay on nature, quoted earlier, Emerson noted that truth derived from nature; however, society was among the many forces that formed the common sense that people needed to pass through the welter of practical demands to achieve the good thoughts they had. Further, contemporary scholars disagree with Dewey's claims that he followed a view of psychology that differed from that of Harris.

As noted in the previous chapter, Dewey asserted that he believed growth derived from physiological factors within the person while Harris considered human growth to be a process in which the person learned to function within the universal system. Despite Dewey's claim, James A. Good argued that the views of Harris and Dewey were roughly similar because they both followed Hegel to frame their ideas. To illustrate his point, Good explained that Harris did not point to a transcendent entity when he referred to Hegel's notion of an absolute or universal system because Hegel had considered the absolute to be something that was not dependent on anything else. Good added that Dewey used the word

DOI: 10.1057/9781137484215.0005

"nature" in a similar manner. Quoting from Dewey's book *Experience and Nature*, Good showed that Dewey viewed nature to be a continuous change that proceeded from a beginning to an end.[3]

Good contended that Dewey's definition of nature was close to Hegel's absolute. He maintained that nature included stones and plants and animals but also politics, myths, and illusions. Good added that Dewey distanced himself from British neo-Hegelians, who described the absolute as a transcendent God or a permanent realm of categories. Nonetheless, Good claimed that Dewey retained a permanent deposit of Hegel's ideas in his own thoughts. This deposit appeared in such concerns as Dewey's view of education as social development, his disapproval of the split between mind and body, and his ideas about the reflex arc in behavioral psychology.[4]

Writing about Dewey's critics, Good described how they made contradictory claims about Dewey's ideas; however, one important complaint was that Dewey was overly optimistic. Citing critics who focused on political issues, Good noted that Reinhold Niebuhr accused Dewey of failing to acknowledge the extent of class conflict, and John McDermott claimed Dewey did not recognize the human capacity to engage in evil deeds.[5] According to Good, critics had raised similar complaints against Hegel.[6]

Critics raised two complaints against Dewey that were important for this essay. First, Amy Gutmann took issue with Dewey's statement in his essay "The School and Society," to the effect that a community should want for all children what the wisest parent wants for his or her child. According to Gutmann, this sentiment revealed feelings of elitism that prevented enlarging the conception of education to make it less individualistic.[7] Second, C. A. Bowers criticized Dewey for aiding environmental destruction by advancing what Bowers called a technological perspective. This was a view that encouraged people to use nature for their own purposes. Bowers felt that Dewey aggravated the difficulties by contending that the scientific mode of experimental thought was the best way to think. Further, Bowers wrote that Dewey dismissed the adoption of sustainable economic practices by describing the members of indigenous cultures who had adopted such methods as savages.[8]

In fairness to Dewey, he thought his idea of a good education could apply to an education that would improve the society by improving the individual. In this way, Dewey believed a good education served both individuals and society. Although Dewey advocated learn by doing, he

DOI: 10.1057/9781137484215.0005

believed this general approach could accommodate several different ways of teaching and learning.[9]

Dewey was not alone in failing to work against the destruction of the environment in the nineteenth century. Although U.S. President Theodore Roosevelt is credited with starting the conservation movement, he did not make it popular until he made his State of the Union Address in 1907. Furthermore, the Roosevelt's plea for a conservation movement was not a proposal to preserve the wilderness as a pristine example of God's creation. Instead, Roosevelt called for legislation to ensure that people would develop the waterways, the timber, the open lands, and the prairies in a manner that allowed future generations to benefit from these resources. The problem was that people erected illegal fences on open lands, exhausted valuable timber, and left soil unprotected from erosion. Roosevelt claimed there was a need for intelligent planning and for laws that enabled people to use the resources in ways to benefit everyone.[10]

Although Dewey was not among the first to recognize the dangers that accompanied the overly rapid settlement of the frontier, his ideas of a good education fit the requirements of the movement for conservation of natural resources when it became popular. The reason was that Dewey applied a technological model of thinking in ways that facilitated the intelligent selection of values. This will become clearer in the next section.

There has been considerable controversy over Dewey's comments when he referred to some peoples as savages. Those remarks in *Democracy and Education* did not concern the abilities or the values of those people. They referred to the limited opportunities for personal growth that was a disadvantage common to all people in traditional societies. The basis of this view was Dewey's belief that people should use as wide a range of possible solutions for problems as they could locate. Unfortunately, the social situation of people living in traditional cultures limited their perspectives, and Dewey pointed to their condition to show how the social medium was educative. Since the children living in indigenous cultures did not attend formal schools separate from daily life, they learned to become contributing members of society by participating directly in adult activities of the society. Although such shared experiences appeared in any society, the modes of life around the children in preindustrial societies required that they spend considerable amounts of time finding food and shelter because they could not control the environment. This limited the range of things they could share.[11]

DOI: 10.1057/9781137484215.0005

From the contemporary perspective, it is unfortunate that Dewey made such references to people living in indigenous cultures. Nonetheless, Thomas D. Fallace devoted an entire book to this question in which he tried to evaluate Dewey's perspective within the attitudes and debates of the period. According to Fallace, those remarks related to a view of psychology built on principles of evolution whereby students worked through activities that illuminated the historical development of human society. Such a perspective implied that human progress was an important element for understanding social arrangements.[12]

Dewey made his strongest complaints against the destruction of the environment after people changed their attitudes about conservation. In this regard, Dewey complained about the ethical lapses that permitted corporations to destroy virgin lands, to pollute rivers, and to waste valuable timber in pursuit of profits. Although Dewey claimed these problems derived from the tendencies Tocqueville had warned against, Dewey's approach built on an effort to turn those problems into benefits. In taking such a position, Dewey depended on the view that such apparent separations as individual and society or stimulus and response were aspects of a wider dialectic. As James A. Good explained, Dewey took this view from Hegel's belief that causes and effects were not distinct entities but existed within relationships that formed complete processes.[13]

As noted earlier, Dewey and Harris wanted the theory of democracy to account for social changes. For Dewey, this meant that educational reforms should relate to the ways social institutions advanced the possibilities of conjoint experiences up to and including the years of World War I. These transitions included the development of science, the expansion of the theory of evolution, and spread of the industrial organization.[14]

At the same time, Dewey expressed the faith in individuals that Harris had developed. John R. Shook quoted Dewey contending that the faith in science and industrial development was not a measure of democracy. Instead, Dewey argued that a democracy could not exist where there was not broad agreement that all people could intelligently solve problems in ways that advanced the communities in which they lived.[15]

For Dewey, education was the way that society could engender the benefits of scientific development, improve the intelligence of the citizens, and enable them to contribute to social progress. The primary way this happened would be for the schools to teach people to think in the same ways that scientists solved problems. Dewey made this suggestion in his descriptions of the workings of the Laboratory School in the University of Chicago, and more theoretically in his book *How We Think*. Dewey wrote the latter book to offer a general principle or theory that could unify the many tasks teachers had to undertake. Dewey argued that teachers should strive to teach students to adopt the scientific method as a habit of thought. He felt this should be easy because the children were animated by a natural sense of curiosity, an active imagination, and a desire to conduct experiments to answer questions. These natural tendencies fit the scientific method of research. If teachers designed lessons to help students practice this method, classrooms would be more pleasant, more productive, and more closely related to life outside the school.[16]

Dewey described "thinking" as a series of steps. The first step was a desire to solve a problem. He did not believe that people started to think until there was some reason inspiring it. Once a person identified a difficulty, the next step was to think of some possible solutions. In order for the person to think in some fruitful manner, these ideas had to come from past experiences. Although other people could make suggestions, there would be no thinking unless the person suspended judgment and tried to determine if those suggestions would work satisfactorily.[17]

More important, the scientific method was not restricted to laboratory settings. Dewey found it to be characteristic of any experience. This was not some activity that simply happened. For Dewey, an experience had two aspects. It involved doing something and undergoing the results. In this way, an experience was similar to an experiment in that the subject learned something. The example he gave was a child placing a finger in a flame. This was the doing. When the child realized that the flame was hot and it burnt the skin, the child had undergone something that he or she could apply to other situations.[18]

In *How We Think*, Dewey did not explain how scientific procedures moved into aesthetic or realms of social value. He dealt with these areas in his major educational text *Democracy and Education*, published in 1916. In the preface, he stated that he wanted to show the educational ideas that would support a democratic society. In the text, Dewey evaluated all arrangements of social life using two measures. The first standard was the existence of numerous and varied interests that the members of the society shared. The second was a full and free interchange among the members of the group with the members of other groups. These measures pointed to a definition of democracy as a mode of associated living or what Dewey

DOI: 10.1057/9781137484215.0005

called conjoint experience. The rationale behind these measures and this definition was that life in a social setting should aim at enabling people to enlarge and enrich their experiences. For Dewey, democracy was the social arrangement best adapted to accomplish this aim.[19]

Since Dewey believed that apparent opposites were often part of the same process, he tried to find benefits that derived from Tocqueville's fears of individualism, conformity, and the desire for material wealth. For example, he noted that individual development could help people have ideas that would improve society, conformity could provide social stability, and the desire for wealth might encourage workers to derive satisfaction from their labor. For Dewey, a proper conception of education would help in this process.

Turning the tendency to individualism toward activities promoting the expansion of people's abilities to contribute to society

As noted in the introduction, one of the problems that Tocqueville felt could weaken the American democracy was individualism. Dewey followed Tocqueville's point and argued that unbridled individualism, which had become a popular view, led to repression; however, he added that it could serve constructive purposes as well.

In an essay entitled "Freedom," Dewey noted that the frontier had made America appear to be the land of opportunity, and it encouraged people to move westward in search of material gain. He added that the situation changed dramatically when free land disappeared and people moved into cities. Nonetheless, the members of the privileged classes continued to believe that they should be able to acquire as much wealth as possible even if they limited the freedoms of other people in doing so. While wealthy people claimed they created an industrial nation, Dewey described how they had destroyed an enormous amount of natural resources, reduced extensive areas of fertile land to arid deserts, and expected the public to combat the resulting floods and erosion. He added that there could not be freedom for people without an abundant store of natural resources. Unfortunately, the courts had accepted claims that economic liberty brought about industrial progress.[20]

Dewey believed that the way out of the problem of individualism was for people to realize the necessity of social life and recognize how

DOI: 10.1057/9781137484215.0005

individuals contributed to its improvement. Dewey argued that the individual was an essential element in the development of knowledge because every new idea originated from a person. In a traditional society, the person who saw things differently appeared dangerous. In a progressive society, the person who perceived the world in unique ways had to be protected because those divergent perceptions led to the discovery of solutions to vexing problems. Dewey believed that progress came when people who thought more deeply than other people created new ideas for everyone to verify and apply. Teachers could prepare students for these roles by allowing intellectual freedom, the free play of diverse thoughts, and the fulfillment of various desires in the classroom. Further, the methods of measuring student achievement had to allow for such differences.[21]

This did not mean that children should ignore social restraints. Dewey pointed out that life involved a continual process of adapting to the changes in the environment, and children were like all living things that had to turn the things that surrounded them to their own use. In nature, the environment might be physical things. In society, the environment included the customs, beliefs, and occupations of the members. Education was the process by which society transmitted these ways of life. Although this could take place through formal or informal ways, Dewey warned that informal education, which depended on children participating directly in essential activities with the adults, was more real than academic training, but it limited the resources available to the children.[22]

The trick was to combine the benefits of direct participation with opportunities to pursue academic training. Dewey tried to bring about such a blending with learn by doing, which will be explained later.

It is important to note that Dewey was not praising people in modern societies who had artificial heat and light, roads, and machines for every purpose when he claimed that people in preindustrial societies could not share a wide range of things because they spent inordinate amounts of time and energy finding food and shelter in an inhospitable, natural environment. Although Dewey recognized that labor saving utilities made it possible for people to expand their stock of knowledge and learn to see the world differently, he did not look upon a catalog of those appliances to represent progress. Those utilities had to function in the interests of a shared social life. That is, people had to control the environment in ways that opened new ways of thinking about social relations.[23]

In a similar way, the wide range of knowledge available in formal education introduced a problem to educators. They had to select materials and activities that enhanced the best mental and moral growth of the students. Dewey recommended using hand work and manual exercises that he called "occupations." These were things the students did that demonstrated the development of society. Since human beings had to acquire food, shelter, and clothing, the occupations explored aspects of these basic drives. For this reason, gardening, weaving, and construction in wood were common occupations in the elementary school.[24]

Froebel's kindergarten predated Dewey's laboratory school. Those teachers gave the name "occupations" to the activities the children undertook. The children used what teachers called "gifts"; these were instruments designed to reveal to the child the unity of things in the universe. For example, kindergarten teachers thought if children sat in a circle, looking at small balls made of wool, they would recall the unity of the universe. In this case, the sitting and looking was the occupation. The balls were the gifts. As might be expected of such an idealistic view, everything was artificial. Other occupations were imaginary performances. The children pretended that grains of sand were seeds they could plant in artificial gardens, and they pretended to pour water on paper flowers. The teachers believed these activities expanded the children's imaginations, enabled them to recall the harmony that existed in the world, and helped them to apprehend the ways people lived and worked together.[25]

Although Dewey followed the pattern of the kindergarten, he turned the artificial occupations into genuine activities that mirrored the industrial development of society. For example, in his school, there were living plants growing in patches of dirt with the aid of real water. The advantage of Dewey's model was that the students learned about the ways various activities developed in society and how the subject matters developed to solve problems that arose in those occupations. From gardening, they learned about agriculture. They learned about the chemistry of the soil and the relation of plants to weather conditions. Most important, they learned about the organization of society that made farming productive. Since the children selected seeds, prepared the soil, and cared for the plants together, they actively participated in a cooperative effort to grow vegetables.[26]

Dewey believed that these occupations transformed the school from a place where students obeyed the commands of the teacher into a miniature community where they had opportunities to share in cooperative activities. The social setting engaged the children's interests as did the

DOI: 10.1057/9781137484215.0005

fact that the occupations were similar to what the children saw outside the school. Most important, these occupations had been essential for human survival. Participating in them would prepare the students to live with others in productive relationships.[27]

The occupations led children through the steps Dewey had listed as constituting an experience. This may seem obvious when applied to practical activities. If a teacher arranged for the students to grow a garden, the work involved doing, such as planting seeds. Although the results implied an undergoing, the educational aim was not the fruits and vegetables from the garden. It was the opportunity to learn how society had developed agriculture. Should the seeds fail to grow, the students would have to look through botany texts to find what conditions the seeds lacked in order to germinate. This could show them what farmers did to protect their crops.[28]

When Dewey pointed out that a failure in gardening could lead to a better understanding of plant life, he suggested that every person's experience influenced the person's subsequent experiences. This applied to a wide range of things. For example, when a child learned to speak, the technical ability of speech opened new opportunities for learning, and this made it possible for the child to have different desires than he or she had before acquiring language. Dewey added that the same thing happened when a society built roads. The experience of constructing highways changed the objective conditions of the environment by offering ease of travel and communication among communities that opened opportunities for people's growth.[29]

Since the product was part of the process, Dewey could apply the same principle of experience to aesthetics. He claimed that genuinely artistic endeavors involved a doing and an undergoing by the artist and the observer. When painters made brush strokes, they engaged in a form of doing. The artists underwent the effects of those strokes when they determined whether the strokes created the picture they wanted. At this time, the artists acted as observant scientists; they recognized cause and effect. Although less obvious, the same relation of doing and undergoing took place in the observer of a painting. In this case, the doing was the observation. The undergoing took place when the observer linked the relations of the parts of the painting to form a recognizable pattern. Viewers who did not form impressions or appreciations of the paintings never went beyond the doing, and they did not have an experience with the painting.[30]

DOI: 10.1057/9781137484215.0005

Because these examples covered a range of activities that were not related to academic achievement, Dewey had to explain what the aim of education was that tied these activities together. His answer was that education did not have an aim. People had aims or desires they wanted to achieve. The personal aims were the products that children made through the examples of educational activities. They were things such as garden vegetables, a dinner with other students feasting on the vegetables they grew, or the painting an artist finished and an observer appreciated. The process was the habits of good thinking they practiced in achieving those aims, and it was the process together with the products that became the aim of education. Dewey pulled these points together to say that if education had to have an aim, it should be growth.[31]

A few years after Dewey wrote the chapter about growth in education, Boyd Bode published a criticism in his book *Fundamentals of Education.* Bode recognized that the ideal of growth offered an escape from the narrowness of other ideals, such as good citizenship or vocational preparation. Nonetheless, Bode complained that the concept of growth was too empty to guide teachers. In fact, he argued that even bad teaching brought about growth. After all, in Fagin's den, Oliver Twist learned to become a pickpocket as he grew in the ability to lift a purse surreptitiously from a person's jacket.[32]

Dewey believed he had taken care of that objection when he wrote that the object of growth was more growth. He returned to the issue in his last educational text, *Education and Experience.* He argued that growth was a condition governed by the principle of continuity. Although people objected that a person could grow as a burglar, Dewey countered that this was not a case of growth. The person may have developed facility at theft, but this was not growth because the ability to steal prevented the person from developing traits that served other lines of endeavor.[33]

The point is that Dewey did not deny the American tendency toward individualism. Instead, he emphasized the benefits it could have when a person expanded his or her talents in ways that contributed to social progress. Furthermore, he defined "progress" as increasing the opportunities for all people to develop their abilities. In these ways, Dewey turned what Tocqueville saw as a condition threatening democracy into a force for its improvement. As will be explained later, it became part of a framework that could encourage environmental protection as well.

DOI: 10.1057/9781137484215.0005

Turning the tendency to conformity to activities revealing the different contributions people made to society

Dewey followed Darwin's theory of evolution to argue that living things controlled the environment for their own purposes. In this regard, he followed the ideas made popular by John Fiske that human beings matured slowly. Fiske claimed the long period of maturation allowed children opportunities to learn to do many different things and inspired people to form societies to help them.[34] When Dewey added that the children could not master the achievements of civilization without the aid of society, he pointed out that this meant that society existed by communication, and it existed in communication. For Dewey, a genuinely social life involved holding ideals, dispositions, and aims in common. While an employer might tell a laborer how to finish a task, their interaction was not social because there was no sharing of interests or purposes.[35]

For Dewey, the way out of the domination that existed between employer and employee was to recognize that shared purposes and interests gave rise to social control in ways that encouraged the fullest development and participation of the members. The example he gave was children's games at school during recess. Such a game might be baseball or football or tag. Each game followed rules and there was some form of arbitration when disputes arose. These two aspects, rules and arbitration, were necessary for the game to continue. They derived from the purpose of the game. In basketball, players had to handle the ball in certain ways or there was no game and confusion prevailed. Dewey suggested that children may imitate adult games they watched or they could follow patterns the children in the school had always followed. The point Dewey made was that the situation controlled the individuals. They shared a common experience. They might play different roles in the game according to their talents. Nonetheless, they contributed to the completion of the game. Most important, it was often the case that no one person was in charge of the activity.[36]

In the case of sports, the aim or purpose of the game appeared within the game itself. A child may role a ball toward another. If the second child rolled the ball back, a game began in which each child modified his or her behavior in accordance with the actions of the child opposite. This observation caused Dewey to note that the children directed their behaviors in hopes that the game would continue, and this was an

DOI: 10.1057/9781137484215.0005

intellectual choice. The individual variation that arose came from differences in children's natural talents while they worked within the accepted patterns. There was a similar set of tendencies in intellectual activities among adults.[37]

As in the case of the children mentioned earlier, an artist was constrained by the materials that he or she used. Dewey pointed out that an artist had to work with canvases, paints, and brushes. The artist had to become aware of the properties of these materials and the ways they responded to different techniques. This body of information represented general methods that the artist had to use, and he or she might devote considerable time and effort to acquire such knowledge. Once the artist had a sense of the general methods of working with the materials, he or she could apply them to new purposes or uses. These innovations were part of what Dewey called the "individual method." According to Dewey, artists who created enduring work applied accepted general methods in individual ways.[38]

Since the individual method appeared when the artist used things everyone knew in a unique way, this required that the artist had to be absorbed in the task on which he or she was working. For Dewey, the best thinking happened when a person focused on the activity at hand. In schools, teachers could distort the process of learning by having the students prepare for tasks they might meet in the future. While Dewey did not oppose preparing for the future, he wanted such preparation to come from fully focused efforts to satisfy present situations. For example, every child would have to use arithmetic, and teachers might assign homework to impart those skills. Dewey argued that the problem in this case was that the future could not stimulate or direct the students' thinking because the homework was not directly connected with the students' present activities. Accordingly, teachers had to provide rewards or punishments, such as grades, to inspire the students to complete the assignments. The students might work in a perfunctory manner, and learning could become a chore to be completed as quickly as possible. A difference would appear when children built kites they wanted to fly. In this case, the nature of the wood, the type of string, and the need for a tail became elements of interest to the child because they were related to the activity. The same was true of arithmetic required to make appropriate measurements in constructing the kite. The task of the child was to use these elements that everyone understood in unique ways to make the kite fly.[39]

The point that Dewey made was that external desires or pressures could distract people from their activities and disrupt the processes involved with learning from experience. For example, Dewey noted that if a farmer cared for animals in hopes of selling them and using the money for something unrelated to farming, every act of feeding or cleaning the stalls would become tasks to be done quickly and with little concern for wider consequences. On the other hand, if the farmer enjoyed being with the animals, the entire range of activities involved in the care of the farm would become significant because each stage in the process provided its own end, and achieving it maintained the activity of farming. In this case, the farmer would be likely to attend to the many aspects of experience that surrounded the task and less likely to harm the environment in an effort to finish the job quickly.[40]

The danger of the pressure for conformity was that it took the person from the activity to some goal that might appear valuable because other people wanted it. This could happen to the farmer who cared for animals to earn enough money to buy clothes of a popular fashion. The drive for the clothes would be the aim and there would be less reason for the activity of caring for the animals to absorb the farmer's thinking.

Dewey did not fight against the pressure for conformity. After all, farmers had to learn from other farmers' experiences what would best help the animals under his or her care. For this reason, conformity or following other examples could be beneficial. The trick was to turn such general methods into individual ones that applied directly to a particular farm and specific animals with an individual farmer. Problems arose when such imitation led to the formation of aims that fell outside the activity itself. In this case, if the farmer thought about the animals and about the things the profit from their sale would buy, one thought would have to distract attention from the other. Schools might help students develop habits to prevent such double mindedness if they reinforced children's natural desire to fulfill their own interests and employed Dewey's model of occupations as lessons.

Turning the drive for material wealth to personal and social satisfaction

As noted earlier, people learned the most from activities in which they had an interest. Another way that Dewey thought about the nature of

DOI: 10.1057/9781137484215.0005

interest was to consider whether the students wanted to solve a problem, which the subject matter could illuminate. In the earlier example, gardening could present several problems that the subject matters of botany and chemistry might help them solve. Unfortunately, when students solved problems in schools, they were the teacher's or the textbook's problems. They were not problems that came from the students. This meant the teacher had to offer a reward or threaten a punishment to encourage the children to undertake the work.

Dewey thought the same process occurred when business people bought large sections of forest even though they had no interest in forest management. They sought a profit. This meant the motive was outside the activity. If the quickest way to gain the profit was to cut down the trees quickly, the owners of the property might have workers clear cut the area even if this practice would prohibit the development of future profits.[41]

Although some entrepreneurs claimed the existence of profits indicated that a business produced useful items, Dewey pointed out that profits did not always depend on honest practices. For example, business owners could artificially restrict the supply of some necessary commodity to gain larger profits than reasonable. Furthermore, advertising created public desires and profits came from satisfying the desires the manufacturers had created. Most important for this essay, Dewey warned that the profit motive encouraged industrialists to choose the easiest and quickest way to produce goods for sale. It led private owners to strip land of mature trees, to rip coal from the ground leaving the property useless, and to exhaust the soil with excessive agriculture to the point where the nutrients washed into the oceans. In these ways, Dewey blamed the profit motive for the waste of natural resources. He recognized that the economic order depended on the profit motive and that it dominated modern life. Nonetheless, Dewey warned that the over-concern for wealth would make intellectual or moral life impossible.[42]

Two years later, Dewey repeated in his book *Art as Experience* that the unnecessary destruction of brooks and green spaces was a product of human greed. In this case, he explained the ways that factories imposed a way of thinking that altered people's aesthetic sensibilities. In making this argument, Dewey was not criticizing science. He complained that factories applied science in a narrow way. They sought to reduce labor and increase profits. This contradicted the correct application of the scientific method because it prevented the selection of the best values from the widest range available.

DOI: 10.1057/9781137484215.0005

First, Dewey argued that science had imposed its own aesthetic through the domination of the machine. As industry grew in importance, skilled workers lost the liberty to turn out objects in pleasing ways. In turn, the machine produced its own aesthetic that consisted of economy of form, and the objects stood apart from people's larger experiences. To Dewey, such objects did not possess beauty although he acknowledged that they had clean, efficient lines. The shift from ornate Pullman railroad cars to steel ones illustrated this transition.[43]

Second, Dewey added that the modern tendency of favoring industry pushed people to ignore the beauty of nature. Although he did not refer to the construction of ugly factories and city slums where forests had grown, he worried that the domination of industry in everyday life had changed people's habits of perception to the point that they tended to disparage the attractiveness of the forms associated with rural life. He thought the traditional love of nature found expression in people's desire to replace the lost natural beauty with things such as flower gardens, but these gardens could not replace the grandeur of oak trees and pine forests.[44]

Third, Dewey repeated that the source of the difficulty was the profit motive. It caused factory owners to limit the application of science in ways that made it impossible for industrial workers to derive satisfaction from their work, caused architects to design tasteless structures with which to fill the cities, and reduced the chances of social progress. As it was, Dewey believed that the drive for profits led land owners to erect shabby buildings and these buildings influenced people's views of the allied arts. They lost their own aesthetic senses because they were surrounded by drab apartments. For Dewey, the way out was to incorporate into the social relations the values that underlay the production and the enjoyment of art. This might encourage some sort of a revolution that would enable men and women to have the freedom to control the processes of production and the capacity to enjoy the fruits of those labors. Accordingly, Dewey felt that art, social justice, and environmental preservation could not prevail until the architecture of the cities where people lived was worthy of a fine civilization.[45]

Conclusion

For Dewey, social reform was possible because societies could grow when people adopted the habits of correct thinking. He did not accept

DOI: 10.1057/9781137484215.0005

the popular view that an accumulation of technical appliances was progress unless those appliances served human growth and development. If the material products were badges of accomplishment, they could prevent people from seeing how technical facility could expand people's activities. He added that those appliances could serve human ends when people applied science as a method of social reform. Dewey believed that the scientific method was the best way of thinking because it enabled people to predict the consequences of some action. As noted earlier, Dewey contended that thinking began with a sense of a problem. It proceeded with an observation of conditions, and the formulation of a possible conclusion. It ended with some effort to test the suggested path of action. Although people could never omit trial and error, he thought the scientific method provided the best way to limit haphazard guessing.[46]

To illustrate the relation between material development and moral improvement, Dewey showed that the ideals of societies could become more humane as the objective conditions of people's lives changed. For example, in ages past, people followed customary morality. Dewey and Tufts pointed out that in ancient Chinese society, if a man beat his mother, the imperial court could order a punishment of death. In addition, the neighbors received beatings and went into exile. Relatives and representatives of the man suffered similar punishments. Such excessive retribution could take place because this was what people had always done; however, as material conditions changed the ideals would improve.[47]

The experience of the ancient Hebrews illustrated the possibility of humane social development. Early records indicated that these people entered Canaan as a warlike, violent people. Their religion described their god as cruel, vengeful, and deceitful. Nonetheless, over time, these people and their religion turned toward justice, kindness, and love. When Dewey and Tufts looked for the reasons for this growth, they attributed it to several factors. Although the factors included the influence of the religious prophets, the expansion of the Hebrew community brought about a need for people to hold to some conception of personal restraint. As codes turned into laws, there had to be ways to forgive transgressions in ways that allowed the community to continue. This led to a reliance on personal spiritual dimensions rather than public ceremonies. Accordingly, the god of Israel became the ruler of the community and people had confidence that this spiritual force would provide for them.

DOI: 10.1057/9781137484215.0005

In this way, Dewey and Tufts argued that the changes in the material conditions of the Hebrew community led to the development of a moral ideal in which justice and love were controlling principles.[48]

Because different conditions could cause different communities to develop different laws, it was not easy to determine what constituted good conduct. Dewey found efforts in this regard among the legacies of ancient Greek philosophers, such as Socrates, who sought the objective bases of morality. Dewey gave the name of reflective morality to similar efforts to find the principles that should direct human conduct in the modern world. When this style of thought was extended, it became moral theory. Although reflective morality sought more objective criteria for correct thinking, it could not provide clear rules for action. Although customary morality provided such reassurances, it could lead to atrocious actions as noted earlier. The traits of reflective morality required that the actor know what he or she is doing. The act must be conscious choice and an expression of a stable character.[49]

Although Dewey warned that reflective morality could not offer clear rules for action, it might help to consider Dewey's strategies of finding what drives caused a problem and seeking ways to turn those impulses toward beneficial results. As noted earlier, Dewey considered the drive for profit and the accumulation of material wealth as overly narrow aims that caused people to overlook other goals such as concerns for the environment. Nonetheless, he saw these material drives as related to a concern for achievement. The alternative that Dewey offered was for educators to encourage students to find satisfaction within constructive activities rather than pursue external rewards such as grades or personal recognition that were unrelated to the activities.

If Dewey is correct, the drive for accountability in schools could contribute to environmental destruction for several reasons. First, dependence on objective standardized tests to determine student success makes every student learn the same thing in the same way. This eliminates the possibility of individual variation essential for social progress. Second, tying such measures of achievement to student promotion and teacher salary raises makes the reason for success external to the activities of learning. In this way, there is a tendency to learn material for the test and little reason to focus on the activities for their own sakes. Third, although science might develop reliable measures of student learning, the narrow application of science would overlook the possibilities of students' need to learn to cooperate and to share in

DOI: 10.1057/9781137484215.0005

activities of mutual interest. Together these tendencies might advance environmental destruction. They would make society less humane.

The solution would be to encourage students to pursue activities in which they have intrinsic interests. While this might mean turning education away from providing training for future jobs, it could mean that schools should provide more satisfying experiences for students. This might make people happier and the environment safer.

Notes

1   John Dewey, *Democracy and Education* (1916, repr., New York: The Free Press, 1944), 87.

2   Michael H. DeArmey and James A. Good, *Origins, the Dialectic, and the Critique of Materialism* (Bristol, England: Thoemmes Press, 2001), 14; Alexis de Tocqueville, *Democracy in America*, trans. Arthur Goldhammer (New York: Library of America, 2004), 595–599.

3   James A. Good, "Rereading Dewey's 'Permanent Hegelian Deposit,' " in *John Dewey's Philosophy of Spirit*, eds John R. Shook and James A. Good (New York: Fordham University Press, 2010), 56–57.

4   Ibid., 57–59.

5   James A. Good, *A Search for Unity in Diversity: The "Permanent Hegelian Deposit" in the Philosophy of John Dewey* (Lanham, CO: Lexington Books, 2006), xvii–xxvii.

6   Ibid., xviii, xxvii.

7   Amy Gutmann, *Democratic Education* (Princeton, NJ: Princeton University Press, 1987), 13.

8   See, for example, C. A. Bowers, "The Insights of Gregory Bateson on the Connections between Language and the Ecological Crisis," *Language and Ecology*, vol. 3, no. 2 (2010): 1–27, www.ecoling.net/download/i/ mark_dl/u/4010223502/4567660582/. accessed 26 September 2013.

9   John Dewey, *The School and Society and the Child and the Curriculum* (1900, repr., Chicago, IL: University of Chicago Press, 1990), 6–29.

10 Theodore Roosevelt's Seventh Annual Message to Congress, 3 December 1907, *Archives of the West*, http://www.pbs.org/weta/thewest/resources/ archives/eight/trconserv.htm, accessed 13 January 2015.

11 John Dewey, *Democracy and Education* (1916, repr., New York: The Free Press, 1944), 16–17.

12 Thomas D. Fallace, *Dewey and the Dilemma of Race: An Intellectual History, 1895–1922* (New York: Teachers College Press, 2011).

13 Good, *A Search for Unity in Diversity*, 202–203.

DOI: 10.1057/9781137484215.0005

14 Dewey, *Democracy and Education*, iii.
15 John R. Shook, "Dewey's Naturalized Philosophy of Spirit and Religion," in *John Dewey's Philosophy of Spirit*, eds John R. Shook and James A. Good (New York: Fordham University Press, 2010), 50–51.
16 John Dewey, *How We Think* (1910, repr., Mineola, NY: Dover Publications, 1997), vii.
17 Ibid., 12–13.
18 Dewey, *Democracy and Education*, 139–140.
19 Ibid., iii, 5, 83–87.
20 John Dewey, "Freedom," *The Collected works of John Dewey, 1882–1953*, Intelex Past Masters, URL: http://pm.nlx.com, accessed 18 September 2014.
21 Dewey, *Democracy and Education*, 291–305.
22 Ibid., 1–3.
23 Ibid., 36–37.
24 Ibid., 196–200.
25 For a description with pictures of Froebel's kindergarten gifts and occupations, readers should see Norman Brosterman, *Inventing Kindergarten* (New York: Harry N. Abrams, 2002).
26 Dewey, *Democracy and Education*, 200–206.
27 Ibid., 346–360.
28 Ibid., 200–201.
29 John Dewey, *Experience and Education* (1938, repr., New York: Touchstone, 1997), 37–39.
30 John Dewey, *Art as Experience* (1934, repr., New York: Capricorn Books, 1958), 5, 44–54.
31 Dewey, *Democracy and Education*, 41–53.
32 Boyd Bode, *Fundamentals of Education* (New York: The Macmillan Co., 1921), 12–13.
33 Dewey, *Experience and Education*, 36–37.
34 See, for example, John Fiske, *Darwinism and Other Essays* (New York: Macmillan and Co., 1879).
35 Dewey, *Democracy and Education*, 3–5.
36 Ibid., 51–60.
37 Ibid., 31–33.
38 Ibid., 170–173.
39 Ibid., 56–57.
40 Ibid., 106–107.
41 John Dewey and James H. Tufts, *Ethics: Revised Edition* (1908; rev., New York: Henry Holt and Co., 1932), 449–458, 487–488.
42 Ibid., 449–458, 447–488.
43 Dewey, *Art as Experience*, 342–343.
44 Ibid., 343.

DOI: 10.1057/9781137484215.0005

45   Ibid., 344.
46   Dewey, *Democracy and Education*, 151, 162–163.
47   Dewey and Tufts, *Ethics*, 15–19.
48   Ibid., 82–99.
49   Ibid., 171–196.

DOI: 10.1057/9781137484215.0005

4

Science, Imagination, and the Environmental Movement: Gregory Bateson's Views

Abstract: *Gregory Bateson complained that scientists caused problems by seeking ways to solve narrow human difficulties. A better approach was for science to use metaphors to help people expand their thinking and illuminate the connections among various parts of the universe. The dangers appeared when people thought they could use the environment for their own purposes. This process caused them to separate themselves from the environment on which their lives depended. This was a moral flaw because it encouraged actions whose effects no one could predict. Such narrow thinking could lead to the destruction of the environment and of humanity.*

Keywords: logical types; metaphor; schismogenesis

Watras, Joseph. *Philosophies of Environmental Education and Democracy: Harris, Dewey, and Bateson on Human Freedoms in Nature.* New York: Palgrave Macmillan, 2015. DOI: 10.1057/9781137484215.0006.

Gregory Bateson used a unique model of thinking to expand the concept of democracy and provide the basis for a new ethical framework that would protect the environment. As noted earlier, Harris thought that democracy prevailed when people had access to important information and opportunities to make their own decisions. He thought schools could prepare students for this notion of democracy if the subject matters introduced them to the truths of civilization, and if they learned that social restraints could enable them to attain high goals. For his part, Dewey thought of democracy as a way of life within which people learned to cooperate to achieve their purposes, and he believed students could learn to do this in classrooms where the curriculum consisted of active occupations in which students followed their interests. Bateson went beyond both of them to expand the sense of community to include the entire environment beyond the citizens that lived there. He warned that when people tried to shape the environment to fit human needs these efforts would lead to the destruction of human life. To predict the effects of such actions, Bateson depended on metaphors rather than on concrete evidence.

Bateson's use of metaphors related to aspects of the scientific method. As the previous chapter suggested, Dewey thought that the scientific method began with some sort of imaginative leap in which a person made a guess as to what might solve a problem or reveal a path to further an experience. For example, in *Democracy and Education*, Dewey noted that the stimulus for thinking was the desire to bring about some consequence. This meant that a person had to imagine the result of some action he or she made. This anticipation suggested a tentative solution, which became the hypothesis, and the trying was the test to see if the guess was correct. These four steps became the process of effective thinking. In various forms, they represented the scientific method.[1]

In *Democracy and Education*, Dewey did not explain how to make the conjectures that could become hypotheses; however, he suggested several ways teachers could help students expand their capacities for imagination. He added that this quality of mind enabled a person to recognize the connection between what existed and what could occur. In addition, he noted that imagination was essential if activities were to become more than mechanical motions. Dewey claimed that imagination brought life to academics. It was the quality of mind that infused meaning into symbols, informed activities, and enabled people to see more possibilities for action within their actions.[2]

DOI: 10.1057/9781137484215.0006

While Dewey suggested that thinking began with a sense of a problem, Gregory Bateson argued that the traditional manner of solving problems caused people to focus on narrow answers. For example, in 1978, complaining that the University of California offered an obsolete education, Bateson explained that difficulties arose because the foundation of knowledge in the University was based on a separation between the ways people thought and any efforts to realize how the world responded to those thoughts. Although Bateson admitted that the university presented the most current facts, it appeared to him that the research to uncover the information overlooked the ways the outer world changed as a result of those human activities. As an example, he pointed to the ways preventative medicine had extended human life. Although this satisfied the demand for greater human comfort, the resulting increase in the population and its many needs caused more difficult problems. According to Bateson, there was no simple way to reduce environmental destruction, yet he suggested that the dilemma posed an opportunity for the board of regents to recognize that the University should recast the purpose of the education it offered. Instead of asking professors and students to solve specific problems, they should ask them to recognize the wider problems or conditions that could result from any proposed solution. This was not an easy task because such a shift would demand extensive reforms to the University. The effort to solve specific problems had trained everyone to apply rigor in selecting the relevant facts and information that would illuminate a possible solution. The effort to predict the effects those changes would have to wider conditions required extensive use of people's imaginations. These patterns of thinking were widely different. The ability to illuminate the nature of a problem did not lead to the facility to predict the changes a proposed solution would cause.[3]

Bateson used an example from Rachel Carson to show how simple solutions created intractable problems. Carson had written about the efforts people took to eradicate mosquitoes with powerful insecticides. Although the aim was to prevent the insects from causing diseases in humans, the insecticides harmed songbirds that ate those insects and allowed the mosquito population to grow. This led to the additional use of poisons that eventually harmed the people the insecticides were meant to protect. Bateson contended that such an effort was immoral because it was based on arrogance. People assumed for themselves the omnipotence they assigned to God. They thought of themselves as created in

DOI: 10.1057/9781137484215.0006

God's image, and they considered the world to be outside the human mind. For these reasons, people exploited nature for their own purposes as if the world did not deserve moral or ethical consideration.[4]

Bateson's complaints influenced scholars in comparative education who called for more research along the lines he described. For example, Victor Kobayashi used his presidential address to the 2007 conference of the Comparative and International Education Society to explain how Bateson's use of metaphors was a promising alternative to the traditional scientific method. Kobayashi claimed that Bateson felt that science sought to control the world of things. This led researchers to separate individuals from their environments, locate verifiable facts, and form general principles. This approach implied that the individual was the unit of survival; however, Kobayashi pointed out that Bateson believed the individual and the environment was the unit of survival. Metaphors could capture the inseparable nature of people in their environments. More important, most people used metaphors without realizing it. In fact, Kobayashi added that Bateson undertook studies that suggested even animals, such as octopi, used metaphors to organize their behaviors. The strength of metaphors was that they operated as aesthetic judgments placing the different elements into an entire system in harmony with each other. For this reason, Bateson sought the metaphors that organized aggregates of facts that people used to think. Although aesthetics implied beauty, some metaphors threatened the life of the community. Accordingly, Kobayashi argued that researchers should identify the metaphors that might lead to mutual destruction of human culture and the natural world when they conducted their studies. Unfortunately, most researchers in comparative education singled out schools in specific countries as their unit of analysis. This eliminated wider comparisons. Accordingly, Kobayashi urged the researchers in his audience to expand their imaginations and seek to illuminate the ways those countries operated within their wider environments.[5]

Bateson's pattern of thinking

By the 1960s, Gregory Bateson had a long list of accomplishments even though he lacked a professional identity or base. He was a pioneer in visual anthropology with his wife, Margaret Mead, and he had worked in areas of psychiatry and communication. According to his daughter,

DOI: 10.1057/9781137484215.0006

Mary Catherine Bateson, the environmental movement gave him a way to integrate the many different disciplines he had employed in the course of those inquiries. The route he chose was to focus on the nature of conscious purpose and the problems that resulted when people tried to change the natural world. As he recognized how parts of the world related to each other, he realized that people followed distorted episte-mologies that led them to interrupt those connections.[6]

Mary Catherine Bateson believed that the environmental movement offered Bateson a way to unify his work in the last half of the twentieth century; however, he had maintained such a unity of purpose since he was a young man. As early as 1941, Bateson wrote that he maintained an unfashionable faith in the unity of earthly phenomena. He credited his father, a famous biologist, for keeping this old-fashioned view alive in him.[7]

To illustrate how all biological systems followed similar patterns, Bateson published an essay explaining what he called Bateson's Rule. This rule was named after Bateson's father, William Bateson, who coined the term, "genetics." Bateson's Rule determined why organisms, such as beetles, grew two appendages in places where they should grow only one.[8]

In his explanation, Gregory Bateson used two metaphors to discuss how cells regenerate an insect's body parts. He drew one metaphor from cybernetics, a type of mathematics used to program such things as guided missiles, and the other from studies of human communication.[9]

From cybernetics, Bateson compared parts of the beetle's leg to a governor, or speed limiter, that kept a steam engine at a constant speed. When the engine turned too rapidly, the governor reduced the amount of fuel available. When the engine moved too slowly, the governor did not operate and allowed the amount of fuel to increase. As for the beetle's leg, Bateson explained that parts of it contained information that prevented other parts from generating an extra foot. When an injury removed from the leg the parts with the necessary information, the absence of information was a message in the same way that the failure of the governor to reduce the fuel allowed the steam engine to increase its speed. Since there was no information preventing the formation of a foot, the cells in the beetle's leg created a second foot.[10]

As evident from this description, Bateson used a metaphor from communication in human society to explain how the beetle's leg knew what to do. That is, he described the insect's body parts as transmitting information to each other so they knew how to grow. In the case of the

DOI: 10.1057/9781137484215.0006

normal beetles, the tissue of the leg surrounding the foot acted like a governor to prevent the body from growing two feet. The unnecessary duplication occurred when an injury to some specimens destroyed the part containing information to block the growth of another foot. This led to the development of the deformity, the extra and unnecessary foot.[11]

When Bateson considered how societies operated, he used a metaphor similar to the process in the beetle's leg. That is, he claimed that society depended on what he called "information." This was the existence of something that made a difference in a later event. For example, when the United States prohibited the legal production of alcoholic drinks, bootleggers began to supply illegal liquor. When the bootleggers made too much alcohol, the police limited the illegal production. Since police could not and did not curtail it, there remained enough alcohol in circulation to keep society in a relatively steady state. Thus, the amount of liquor available in society was the information that influenced the extent of the efforts of law enforcement agencies.[12]

To make such metaphors, Bateson benefitted from several influences. As noted earlier, he credited the close association he had with his accomplished and intellectual father. In addition, he profited from close association with gifted teachers and colleagues who helped him arrange his discoveries about human nature to illuminate evolution, social change, and environmental protection.

When Bateson entered Cambridge University, he began studying biology. Upon finishing this course, he went on to study anthropology. His professor A. C. Haddon sent him to New Guinea in the 1930s to live with the Iatmul and to report on their ceremony called the "naven." This was the celebration for a person who had performed a culturally significant act. Since the Iatmul were head-hunters, one such important event was killing a member of another tribe. He approached the work with the belief that he could find the interlocking relationships of behaviors that made up the Iatmul culture. While he was working in New Guinea, he obtained a copy Ruth Benedict's book *Patterns of Culture*, and it expanded his thinking to look for rules determining the complex social organization of these people.[13]

Two theories arose out of Bateson's research in New Guinea that led him to construct theories about the social influences on human behavior. The first theory was schismogenesis, and the second was eidos. Schismogenesis referred to the tendencies of people to direct their actions according to the responses of other people. The theory of eidos

DOI: 10.1057/9781137484215.0006

was the tendency of the culture to direct individuals to develop standard or approved ways of thinking that the society selected from the store of attitudes or values the members selected. Among the Iatmul, they could recite long lists of names or totems, but they seemed less able to recall the sequence of events that took place in some situation. Bateson referred to these tendencies as cultural elements within the minds of the individuals that led to patterns of human behavior.[14]

The metaphor that Bateson used to determine the influence of schismo-genesis came after he returned from New Guinea and discussed European politics with Alan Barlow before the onset of World War II. Industrialized nations built increasingly powerful weapons to maintain parity as they prepared for war with each other. The events during the ceremony of the naven seemed to follow this pattern. As the men pretended to be women, the women reacted enthusiastically. When the women exaggerated their responses, the men became more exhibitionistic and exaggerated their actions. Bateson coined the term to mean the creation of divisions. Later, it appeared to him that such symmetrical or competitive relationships appeared in the arms race during the Cold War.[15]

On the one hand, Bateson used the metaphor of a military arms race to determine the process by which he could describe and predict the behavior of the men and the women during the naven. From this metaphor, he translated the phrase, the creation of divisions, into Latin to coin the term "schismogenesis." On the other hand, the concept of schismogenesis is a metaphor because the concept or theory exists in the mind of the anthropologist and he or she must compare the group's behavior to the theory to gain deeper insights into what might happen. The result was that Bateson found himself using empirical evidence to show the relevance of a metaphor more than he could use it to prove something correct. Nonetheless, with such metaphors, he could suggest the possible results of any action, including the effects of a proposed solution to a problem.[16]

The reason metaphors were not open to proof was because metaphors compared things without spelling out the comparison. For example, Bateson noted that people would say a nation decays, but they would not show how the process of decay that applies to fruits could apply to nations. In a similar way, Bateson believed that when male animals approached females of the same species, they acted out metaphors of the relationship of parents and children without specifically noticing the similarities.[17]

DOI: 10.1057/9781137484215.0006

Turning the dangers of individualism toward cooperation by creating metaphors that illuminated the social influences on individual thinking

World War II began after Bateson and Margaret Mead returned from their research in New Guinea and Bali. Living in New York, they participated in a seminar on the problems of preserving the democratic ideal of individual freedom during the quest for security while the war was on. Noting that authorities could teach children to report if their parents planned any treacherous actions, Bateson extended the notion of eidos to show that such an effort would weaken the democratic ideal. According to Bateson, the problem came from the ways people learned to learn. That is, they would pick certain aspects to make standard from the fund of available values. Bateson coined the term "deutero-learning" to locate the tendency of an organism to become more skilled in solving problems by solving problems. The awkward phrase, deutero-learning, implied that there were different types of learning taking place. The arrangement of the various types of learning appeared in a hierarchy. Accordingly, solving simple tasks was one type of learning but becoming adept at learning simple tasks was another. On the hierarchy, the simple response of an animal to a stimulus was at the lowest level. An animal reached another level when it realized that certain behaviors, such as hitting a button, produced specific results, the release of food. Bateson went on to explain that, in the case of teaching children to report their parents' actions to authorities, the lessons went beyond those simple limits. When patriotism or official state mandates appeared superior to family ties, the lessons would change the children's perception of their parents and thereby weaken the fabric of society. Bateson ended his essay with a comparison to Balinese society, which he and Mead had just witnessed. Those people organized their actions through ritual courtesy that they believed would ward off an unnamed danger lurking over them. Bateson suggested that Americans should cultivate the opposite tendency. They might live in the hope that they could raise children to grow into happy adults by reinforcing the relationships among parents and children.[18]

In the earlier example, Bateson joined two metaphors to make his prediction about the dangerous effects of a proposed policy. The first metaphor was the notion of eidos. The authorities who wanted children to report the actions of their parents would standardize the appropriate values children should fulfill around patriotism. The other metaphor was

DOI: 10.1057/9781137484215.0006

the notion of deutero-learning, which spelled out the various alternative values the society could expect children to internalize. The prediction of baleful effects derived from the children shifting their allegiance from their parents to the state.

Two examples illustrate how Bateson showed the ways the concept of individualism and the belief that a person can control the environment led to problems. The first came in the 1950s when Bateson and several colleagues worked in mental hospital devising a theory of what was called "schizophrenia." The second was when Bateson developed a theory about alcoholism.

To develop a theory of schizophrenia, Bateson and his colleagues called upon metaphors about levels of communication and upon the theory of logical types. The idea of levels of communication came from his work in New Guinea. When Bateson described the ceremony of naven, he postulated different levels of communication with his notions of schismogenesis and eidos. These implied that the reactions of people who witnessed some behavior made up one level of behavior. The relevant social standards created another level. Each of these levels changed the meaning of the behavior. At the same time, Bateson and his colleagues used the theory of logical types to complete the meaning of these explanations.

Bertrand Russell and Alfred North Whitehead developed the theory of logical types. Although they did not know how far it stretched, Bateson simplified the theory of logical types to the point where he claimed that it meant the context of an event or of a communication operated by different rules than the content. For example, if a person smiled while issuing a command, he or she could transform it into a pleasant request.[19]

In the case of the theory of schizophrenia, the context was the means of classifying the message in one of several modes. Bateson thought the modes of communication included such things as play, fantasy, or humor. Although the idea of logical types suggested that a person could change the meaning of an insult by assuring the hearer that it was a humorous expression, these contexts could turn toward malevolence. One such example took place when the mother of a patient at a mental hospital visited her son. The patient had shown improvement. When his mother appeared, he appeared delighted to see her and reached to embrace her. She stiffened. He withdrew his arms. She asked why he withdrew his embrace if he loved her. After the mother left, the patient struck a guard violently.[20]

DOI: 10.1057/9781137484215.0006

Most people would refuse to fall into the mother's trap. The patient had tried to express affection, and the mother had rejected it. Then, she denied having refused it, and she criticized the patient for not offering genuine love. In such a case, most people would complain to the mother that she was the culprit; however, the patient had a long relationship with his mother during which he had learned that he could never criticize her. In this relationship, the mother placed the patient repeatedly into similar situations where all his actions were wrong and she would be justified in withdrawing her love for him. Bateson believed the patient fought with a guard to relieve his frustration at being trapped in such a double bind.[21]

Subsequent therapy with the patient's mother revealed the possible truth of this description of the double bind. Bateson found that the mother disliked her son because he was the product of an illicit relationship. Worse, the mother was locked into competition with the members of her family. When the mother was a little girl, her grandmother had thrown a knife at her and nearly harmed her. The mother had retaliated some time later by throwing objects at her grandmother and her mother. In this way, she dominated them and this made her feel superior to them. Nonetheless, the mother felt a strong need for her family to love her. Although she transferred the tendency to dominate people to her son, she needed his demonstrations of affection to reinforce her belief that he appreciated her care and that she was loveable. Bateson argued that when the mother told her son that he should express his affection even when she showed disdain, she defined him as unable to express normal emotions and offered justification for her to sever her relationship with him. Trapped in such a pathological relationship, the son responded by acting violently.[22]

The point in these examples is that Bateson was able to show how individuals could not think independently. Their perceptions were influenced by social influences they could not recognize unless they looked well beyond themselves. While Freudian psychologists would look for internal impulses such as sex drives or desires for power, Bateson settled on more external relationships as symbolized in schismogenesis, which looked at the influences of other people's responses, and in eidos, which was the tendency of the culture to influence the individual's ways of thinking. It seemed reasonable that if people were influenced by the people with whom they lived, they could improve their own behavior by choosing to live with people who followed metaphors that derived from an ethical framework directing them to cooperate in serving the common good.

DOI: 10.1057/9781137484215.0006

Reducing the problems of conformity by recognizing the dangers of inadequate metaphors

The second example of the dangers implicit in the notion of individualism and the effort to control nature came from Bateson's explanation of the problem of alcoholism. Drawing his analysis from the apparent success of the organization Alcoholics Anonymous in counteracting the addiction, Bateson contended the frequent lapses into drunkenness that alcoholics reported suggested the addicted person was trapped in a cybernetic or self-governing system. According to Bateson, drinking alcohol was relaxing when it took place within a convivial setting with complementary relations among the participants. Problems arose when the addicted alcoholic asserted his or her individualism by competing with the drinking partners to drink more than they did. The competition did not have to be overt. It could be a subconscious effort to match other people drink for drink. The problem was that such a contest turned complementary relationships into symmetrical ones in which the individual tried to assert his or her superiority. The same metaphor continued once the person was addicted. The person suffering from alcoholism tried to defeat the bottle. Unfortunately, the bottle retained its attractiveness while the individual fought the urge to drink. The fatiguing nature of the symmetrical relationship with the bottle would lead the addicted individual to recall the context of conviviality that the bottle had represented before he or she tried to beat it. The addicted person surrendered and began to drink again, which led to the battle all over again, lapsing into reconciliation, and leading to surrender complete with a return to drunkenness. Bateson thought that the success of Alcoholics Anonymous derived from its requirement that the addicted person acknowledge his or her powerlessness in the face of alcohol. This admission restored a sense of complementarity to life and rejected the competitive symmetry of overcoming the bottle. The metaphors changed from being deadly forms of competition to being life giving in the acceptance of powerlessness.[23]

Bateson made an important point in the letter he wrote to the regents at the University of California. As noted earlier, Bateson warned that the aim of solving problems allowed people to ignore the responses that the world made to such efforts. In this way, people thought they could gain power over nature through scientific research, but this ideal was a myth. It placed human beings in opposition to nature. The problem was not so

DOI: 10.1057/9781137484215.0006

much that scientists could not subjugate nature as the alcoholic could not overcome the bottle. The problem was that the myth of power corrupted people in the same way the pride of the alcoholic insisted that all that was lacking was the will power to resist temptation. Bateson called inadequate theories showing people suffering from alcoholism as immature, maternally fixated, or weak. They missed the point that the dependency on alcohol derived from a corrupting metaphor. They suffered the sin of pride.[24]

When he turned to explain people's actions, Bateson used the idea of logical types as a controlling metaphor. Taking the act of chopping down a tree, Bateson disagreed with the view that a person did something to a tree even though most people separated each entity so that the person, the tree, and the ax were three separate bodies. For Bateson, a better model was to make the mind appear as a cybernetic system divided into levels. Linked together, the levels provided an ecological frame wherein the mind was part of nature and inseparable from it. The wood chopper gauged his or her actions through differences in nerves and muscles as well as changes in the tree's stump and the flight of the ax. These formed circuits through which the person modified his or her actions to suit the task.[25]

Although Bateson contended that human beings sought to integrate their levels or types of knowledge, he added that many of the levels of knowing were unconscious. For example, people could describe what they saw, but they could not explain how the mechanisms within their bodies provided them with those images. Bateson believed that most people ignored this inability because conscious thought isolated aspects of life as people tried to solve problems; however, he believed that art was fundamentally different from conscious thought. According to Bateson, the artistic impulse came from an effort to communicate those unconscious patterns of knowing. In addition, the artistic impulse derived from a desire to show that those patterns were part of the external world. Accordingly, art could correct the tendency to separate the parts of life from each other.[26]

In fact, Bateson noted that some people seemed to have an innate ability that permitted them to work within complex interacting systems rather than separate them as a scientist would. On the one hand, Bateson considered such an innate ability to be an aesthetic sense. On the other hand, he thought of the ability to recognize classes or types of systems to be essential for moral judgments. For this reason, he thought skilled practitioners, such as baseball players, could help him understand the

DOI: 10.1057/9781137484215.0006

way this aesthetic sense worked. It seemed to him that baseball players did not break everything into constituent parts. Instead, they used some sort of central metaphor to guide their actions. Although some people may not consider baseball to be an art, Bateson defined the ball player's ability as an aesthetic one.[27]

It is important that Bateson derived his ideas of aesthetics from a study of what was called "primitive art." This was the painting or sculpture of preindustrial societies. The important question for Bateson was how critics raised in one culture could recognize the meaning of art from other totally different cultures. For example, the figures from ancient African and Oceanic civilizations influenced Pablo Picasso as he created such paintings as Les Demoiselles d'Avignon. According to Bateson, the reason such aesthetic principles could cross cultural boundaries was that they integrated aspects of knowing that otherwise remained separate. For example, in painting, the artist had to control the materials with such dexterity that the technique does not interfere with forming the patterns in the picture. Bateson called such integration "grace." To demonstrate this quality, Bateson showed a picture painted in 1937 in Bali. For example, the foliage was stylized according to the methods practiced by other Balinese painters. These were measures of quality among Balinese painters. Nonetheless, the painting was complex in that it contained references to many aspects of life: serenity, sex, birth, and death. This led Bateson to conclude that it was an effort to integrate the many diverse aspects of life in one picture, and had the artist focused on any one aspect it would have resulted in the error of omitting another aspect.[28]

While primitive societies respected the different aspects of life, Bateson felt that modern society moved in a direction to make everything similar. To support the needs of the residents of cities, humans created single specie fields of corn or factory farms for poultry, and they proposed sweeping corrective for minor problems, such as weed control. The enclosure of poultry in confined areas posed the problems of pollution and contagious diseases. Blight could sweep through a huge area of cornfields and reduce the yield. Farmers using systemic herbicides could destroy millions of honey bees. Nonetheless, Bateson thought some correctives existed. For example, human love could avoid the narrowness of human purpose. In this regard, Bateson pointed to Martin Buber and suggested that people had to form I-Thou relationships with their ecosystems rather than I-It relationships in which people used the environment.[29]

DOI: 10.1057/9781137484215.0006

Using metaphors to move a normatively correct pattern of thought from philosophy to natural history

Some anthropologists complained that Bateson spent time thinking about his thinking instead of collecting information. For example, Michael Houseman and Carlo Severi compared Bateson's accounts of the Iatmul ceremony, naven, with the results of other researchers. They claimed that Bateson's theoretical arguments did not match the evidence from other anthropologists.[30]

Bateson predicted this problem would occur. Acknowledging that the book *Naven* was clumsy, he claimed the problems resulted from the fact that three levels of abstraction appeared in it. Nonetheless, he added that this was the way science could advance. The first level was the ethnographic information about the Iatmul. The second level was an effort to fit the data together in a picture of the culture, and the third was the result of an inspiration that Bateson had as publication neared. He realized that his theoretical constructions were only his way of putting the puzzle together. They were descriptions of processes of knowing. To imagine that ethos or social structure had reality was to make the error of misplaced concreteness. Social structure does not influence people. Instead, it is a class of explanations of things that could determine a person's behavior.[31]

Bateson went on to explain that this train of thought had come to him as he taught cultural anthropology to aspiring psychiatrists. In the course of these classes, he realized that he had written *Naven* without reference to Freudian thought even though the Iatmul symbolism was rife with sexual connotations. Nonetheless, Bateson thought it fortunate he had overlooked these possibilities because it may have prevented him from developing the concept of schismogenesis. As noted earlier, schismogenesis was the tendency of people to direct their action according to the responses of other people. Bateson suggested that the concept of schismogenesis might offer better explanations for several phenomena. It seemed to explain some things about the ways people formed their characters. Bateson went further afield suggesting the schismogenesis might control the direction of evolution. It seemed to him that popular theories of evolution suggested that the process took place haphazardly. Although this was termed "natural selection," there was no sense of direction implied by this term. Searching for some explanatory principle, Bateson wondered if biological systems worked together so that the

DOI: 10.1057/9781137484215.0006

proximity of other creatures influenced the direction in which organisms changed.[32]

In explaining the direction of growth, even evolution, as resulting from interactive processes, such as schismogenesis, Bateson reaffirmed his rejection of dualisms. Natural selection seemed to place the environment in opposition to the organism. He rejected any consideration of God directing nature because this implied a separation between the ruler and the ruled. Instead, Bateson saw all the entities in the world in relation to each other so that changes in one led to reciprocal changes in others. He claimed this pointed to mind as the explanatory principle for change. Accordingly, he rejected theories that depended exclusively on explanations of material change. In this regard, he claimed to follow Samuel Butler in rejecting Darwinian evolution because it avoided any sense of what he called "mind."[33]

Bateson acknowledged that he began pulling together his conception of thinking with his observations in New Guinea where he developed the notion of schismogenesis; however, his experiences with Adelbert Ames, Jr., who had created experiments involving perception, helped him realize that his mind created the images of the world that he saw. From these experiences, he came to think of mind as a property of natural history.[34]

Ames's experiment involved optical illusions. The subject had to move a plank through a series of levers toward a package of cigarettes placed on a spike three feet away and a book of matches placed on another spike six feet away. Before moving anything, the subject looked at the arrangement from above and saw the objects as they were. Second, the subject had to look at the objects with only one eye, through a hole, set at the level of the table. When the subject moved a plank sideways, the position of the objects changed, and their relative size seemed to reverse. The cigarettes grew and the matches shrank although, in fact, these changes had not taken place.[35]

The point Bateson drew from this experiment was that he had created an image of a scene from the multitude of impulses that fell on his optic nerve, and that scene appeared differently than it was. He did this unconsciously by applying the rules of parallax that he had acquired from earlier experiences even though those rules were not appropriate for the situation. To Bateson, this meant that people saw what they created without realizing the act of creation.[36]

At this point, Bateson had two problems to solve. His mind seemed to create the image he perceived without his being aware of how it

DOI: 10.1057/9781137484215.0006

happened. This left him with the sensation that the viewer and the world were separate even though he knew intellectually that this was not true. He claimed he solved the riddle by deciding that reality may be outside the person, and perception may be inside the person, but the mind made sense of the sensations by recognizing differences in the stimuli it received. Since the differences in the world made differences in the person's mind, it appeared to him that the mind was partly in both places. Most important, since he thought by means of differences, he could stop thinking of the world as a collection of material entities outside his mind. Most important, the fact that there seemed to be some sorts of feedback in the universe established its permanence. This left him with the need to account for the problems of levels because the stimuli were of differences and of differences among differences.[37]

He solved the issue about the levels of learning through the theory of logical types. As noted earlier, this was the view the members of a class operate by different rules than the classes work among themselves. In this case, one level of learning was about a difference and a higher level was about a difference among differences. For example, Learning I was similar to the conditioning to which Pavlov subjected his dogs. In this case, the dogs associated the sound of a bell with food; when the bell rang the dogs salivated. Learning II represented deutero-learning in which the subject was able to change the process of learning. For example, authorities might condition children to report their parents' acts of treachery; however, Learning II occurred when the children recognized that those authorities did not help families. Learning III represented a change in the set of alternatives for learning. For example, in Learning II, the person learned something about the context of Learning I. In Learning III, the person learned about the context of Learning II.[38]

Bateson thought that this epistemology had changed the search for truth from being philosophical to being part of the realm of natural history because he relied on descriptions of the sensory machinery people used to learn about things in the world. The normative aspects derived from the fact that people made errors when they created separations between thoughts and objects in the world. Bateson had done this by looking at schismogenesis, double-bind theories of schizophrenia, and the theory of logical types. Unfortunately, human languages depended on words and parts of speech that divided thoughts about the world and made things appear opposed to each other. When people assumed that language correctly described the world, they confused a map with reality. The map was a representation; it was not the world.[39]

Bateson considered his epistemology to be normative in two ways that it corrected the errors people made. First, since language tended to present the world as composed of separate parts described by nouns and verbs, Bateson pointed out that people would be more accurate to speak of levels of difference, as in the case described earlier of a person chopping a tree. In this way of thinking, the entities involved, the person, the ax, and the tree, were parts of a unified system. Second, a more pragmatically serious error resulted when people separated themselves from the objects they used and the things they did. People made this error by considering their thoughts to be their own and the objects in the world to be mindless things they could exploit. Such errors made it appear reasonable for people to see the outside world as something they could control, and there was no limit to how far they could go in using the world. This was a cause of environmental destruction. Bateson feared this error would lead to the end of human life because he considered the unit of survival to be the biological specie and its environment.[40]

Conclusion

Although Bateson could not explain how a person could move from one level or type of thinking to another, this type of conversion seemed to be the required answer to the ecological crisis. In the case of the alcoholics, described earlier, they had to reduce themselves to the point where they could not retain any confidence in their ability to end drinking on their own. At this point, they could realize their pride was the problem. Among alcoholics, this was called hitting the bottom, and it might not ever come during the alcoholic's life. If this description of alcoholics was a metaphor for human civilization and its environment, the situation looked very dim.

Furthermore, it was so difficult to create an epistemology without dualisms that Bateson worried he may have failed. For example, during a symposium on human adaptation, Bateson explained that he warned against thinking the solution to problems was to make some alteration in the world. As in Rachel Carson's criticism of using DDT to eradicate mosquitoes, described earlier, the difficulties arose from not recognizing the relationships among the mosquitoes, songbirds, and the insecticide.

DOI: 10.1057/9781137484215.0006

Since these oversights happened frequently, they appeared to be willful and this placed them in a category of moral concerns. Unfortunately, Bateson came to think that he suffered from similar oversights when he criticized the technological search for technological solutions to technological problems. This type of search was deeply rooted in Western culture, and he did not know what would happen if people tried to root out the motives for such ways of thinking. For this reason, he feared he was guilty of acting without a clear conception of the results.[41]

Because he might be guilty of arrogance, Bateson offered modest reforms for contemporary society when he was asked how people should restructure a city. An important point he made was to warn people to avoid trying to return to the innocence of pre-industrial, indigenous people. He believed that such romantic efforts would destroy the wisdom that prompted the return. Second, he called for the use of computers and communication devices that would enhance the physical, aesthetic, and creative lives of the people. Although he approved of retaining modern technology, he thought people should limit their consumption of goods to the point where they took from the natural world only what they needed for necessary change. To accomplish this goal, he urged people to retain flexibility within themselves, their society, and their environment. This meant that the legal system should not be overly restrictive, and the cultural premises should be as flexible as possible. People should be encouraged to use their freedoms. Most important, he urged that people learn to think ecologically because such understanding was more important than obeying maxims about preventing pollution.[42]

The last point implies that human wisdom is the most important tool in preserving the environment. At the same time, this suggestion will preserve human life because the possibility of achieving wisdom is the quality that defines human beings as human. Readers may recall that Bateson offered a different solution when he urged people to develop I-Thou relationships with the environment. This represents another metaphor, because Bateson compared his notion of a naturalistic epistemology to a philosophical one without describing the ways they could merge.

The possibilities of mixing two incompatible epistemologies may not be as far-fetched as it sounds. In fact, Martin Buber tried to describe such a blend. He wrote that reality for people was meeting, yet they were torn between two incompatible types of relationships. On the one hand, the I-Thou relationship, in which the Thou and the I were truly present

DOI: 10.1057/9781137484215.0006

to each other, was where people were fully human. Within this bond, neither side recognized the individual characteristics or the possible usefulness of the other side. Unfortunately, people could not live in a spiritual world forever. To remain alive, people had to turn objects of devotion into things they could use. For example, although husbands and wives loved each other fully, they appeared to each other as help mates as well. Buber called this relationship of mutual usefulness an I-It relationship. In this relationship, people recognized specific aspects of each other, and they saw ways to use each other. For these reasons, Buber said that people were trapped in a tragic condition. They could not be human unless they had I-Thou relationships, but they could not live by love alone. I-Thou relationships had to become I-It relationships.[43]

Bateson might say of people's relation with the environment what Buber said of people's relationships generally. If the outside world was part of mind, people could not live unless they loved nature without reservations; however, they could not live by means of such love alone. At times, they needed to take from nature to provide for their livelihood. This may be why Bateson recommended that people act within strict limits taking no more from the environment than they needed for necessary change. At the same time, they should continue to learn and grow. This might be a reasonable beginning for an ethical framework that would preserve the environment.

Notes

1   John Dewey, *Democracy and Education: An Introduction to the Philosophy of Education* (1944 repr., New York: The Free Press, 1916), 151.

2   Ibid., 236–237.

3   Gregory Bateson, "Appendix: Time is Out of Joint," in *Mind and Nature: A Necessary Unity,* ed. Gregory Bateson (Cresskill, NJ: Hampton Press, 2001), 201–210.

4   Gregory Bateson, "Form, Substance, and Difference," in *Steps to an Ecology of Mind* (Chicago: University of Chicago Press, 1972), 454–471; Gregory Bateson, "Style, Grace, and Information in Primitive Art," in *Steps to an Ecology of Mind* (Chicago: University of Chicago Press, 1972), 128–152.

5   Victor Kobayashi, "Recursive Patterns that Engage and Disengage: Comparative Education, Research, and Practice," *Comparative Education Review,* vol. 51, no. 3 (August 2007): 261–280, Stable URL: http://www.jstor. org/stable/10.1086/518464, accessed 7 January 2015.

DOI: 10.1057/9781137484215.0006

6   Mary Catherine Bateson, "Foreword," in *Steps to an Ecology of Mind*, ed.
     Gregory Bateson (Chicago: University of Chicago Press, 1972), vii–xv.

7   Gregory Bateson, "Experiments in Thinking about Observed Ethnological
     Material," in *Steps to an Ecology of the Mind* (Chicago: University of Chicago
     Press, 1972), 73–87.

8   Gregory Bateson, "A Re-examination of Bateson's Rule," in *Steps to an Ecology
     of Mind* (repr., 2000, Chicago: University of Chicago Press, 1972), 379–399.

9   Ibid., 379–399.

10   Ibid.

11   Ibid.

12   Gregory Bateson, "Effects of Conscious Purpose on Human Adaptation,"
     in *Steps to an Ecology of Mind*, ed. Gregory Bateson (2000, repr., Chicago:
     University of Chicago Press, 1972), 446–453.

13   Gregory Bateson, *Naven*, 2nd edition (Stanford, CA: Stanford University
     Press, 1958), 1–10.

14   Ibid., 222–225, 280–303.

15   Ibid., 177, 218–220, 280–302.

16   Gregory Bateson, "Culture Contact and Schismogenesis," in *Steps to an
     Ecology of Mind* (Chicago: University of Chicago Press, 1972), 61–72.

17   Gregory Bateson, "Metalogue: What Is an Instinct," in *Steps to an Ecology of
     Mind* (Chicago: University of Chicago Press, 1972), 38–58.

18   Bateson, *Naven*, vii; Gregory Bateson, "Social Planning and the Concept of
     Deutero-Learning," in *Steps to an Ecology of Mind* (Chicago: University of
     Chicago Press, 1972), 159–176.

19   Bertrand Russell, *Introduction to Mathematical Philosophy* (1993, repr., New
     York: Dover Publications, 1919), 135–138; Gregory Bateson, "Minimal
     Requirements for a Theory of Schizophrenia," in *Steps to an Ecology of Mind*
     (Chicago: University of Chicago Press, 1972), 244–269.

20   Gregory Bateson, "Toward a Theory of Schizophrenia," in *Steps to an Ecology
     of Mind* ed. Gregory Bateson (2000, repr., Chicago: University of Chicago
     Press, 1972), 201–227.

21   Ibid.

22   Ibid.

23   Gregory Bateson, "The Cybernetics of 'Self,'" in *Steps to an Ecology of Mind* ed.
     Gregory Bateson (2000, repr., Chicago: University of Chicago Press, 1972),
     309–337.

24   Bateson, "Appendix: Time is Out of Joint," 201–210.

25   Bateson, "Form, Substance, and Difference," 454–471.

26   Bateson, "Style, Grace, and Information in Primitive Art," 128–152.

27   Bateson, "The Moral and Aesthetic Structure," 253–257.

28   Bateson, "Style, Grace, and Information in Primitive Art," 128–152.

29   Bateson, "Effects of Conscious Purpose," 446–453.

DOI: 10.1057/9781137484215.0006

30   Michael Houseman and Carlo Severi, *Naven of the Other Self: Relational Approach to Ritual Action* (Boston: Brill, 1998).

31   Ibid., 281–283.

32   Ibid.

33   Gregory Bateson, "This Normative Natural History called Epistemology," in *A Sacred Unity: Further Steps to an Ecology of Mind* (New York: A Cornelia and Michael Bessie Book, 1991), 215–229.

34   Ibid.

35   Gregory Bateson, "Pathologies of Epistemology," in *Steps to an Ecology of Mind* (Chicago: University of Chicago Press, 1972), 486–495.

36   Bateson, "This Normative Natural History," 215–229.

37   Ibid.

38   Gregory Bateson, "Logical Categories of Learning and Communication," in *Steps to an Ecology of Mind* (Chicago: University of Chicago Press, 1972), 279–308.

39   Bateson, "This Normative Natural History," 215–229.

40   Bateson, "Form, Substance, and Difference," 454–471.

41   Gregory Bateson, "The Moral and Aesthetic Structure of Human Adaptation," in *A Sacred Unity: Further Steps to an Ecology of Mind* (New York: A Cornelia and Michael Bessie Book, 1991), 253–257.

42   Gregory Bateson, "Ecology and Flexibility in Urban Civilization," in *Steps to an Ecology of Mind* (Chicago: University of Chicago Press, 1972), 502–513.

43   Martin Buber, *I and Thou*, 2nd edition, trans. Ronald Gregor Smith (New York: Charles Scribner's Sons, 1958), 3–18.

DOI: 10.1057/9781137484215.0006

Where Do We Go from Here?

Abstract: *The conclusion shows that the best environmental education takes place when teachers seek to expand students' understanding of the world. There are a range of practical solutions for environmental destruction. They include ending capitalism, creating more parks, or blocking further immigration. Unfortunately, each of these suggestions would increase the abuse of the environment. Unless the tendencies for individualism, materialism, and conformity diminish, the environmental problems will escalate.*

Keywords: park conservancies; problem posing education; world views

Watras, Joseph. *Philosophies of Environmental Education and Democracy: Harris, Dewey, and Bateson on Human Freedoms in Nature.* New York: Palgrave Macmillan, 2015. DOI: 10.1057/9781137484215.0007.

DOI: 10.1057/9781137484215.0007

This book selected philosophers who described a moral framework within which a democratic ethic could operate to serve the common good. As Tocqueville noted, democracy released tendencies of individualism, materialism, and conformity. These could serve beneficial as well as dangerous tendencies. The important element was to find ways to direct those tendencies so that individual freedom advanced the common good. The philosophers discussed in this book worked around that apparent contradiction by constructing moral frameworks based upon the recognition of the connections among the things in the world.

As noted earlier, Harris wanted students to recognize that the constraints of social institutions, such as the family, civil society, and the church, enhanced their freedoms. To create such a curriculum, Harris adopted Hegel's approach of integrating all elements of human experience into a comprehensive system that reflected the development of human society. Dewey absorbed the lessons Harris offered, but he turned the Hegelian drive to unify human knowledge into a belief that science offered the best way to solve problems. Although such an effort could lead to a formulaic approach to life, Dewey thought that the scientific method opened human thought to increasingly wider and more socially beneficial efforts. For this reason, he described students who began by making bows and arrows, but trying to improve those implements led them to create a forge and make iron. Thus, a pedestrian impulse became the inspiration for an experiment in metallurgy.[1]

While Harris and Dewey built on Hegel's system of logic, Bateson expanded an essential element of the scientific method to help people understand the problems they caused when they tried to control nature. The element of thought that Bateson expanded was imagination. Scientists needed this quality of mind to form hypotheses. In framing a hypothetical explanation of the solution to a problem, people needed to imagine what would happen when something was done. In his text *Democracy and Education*, Dewey acknowledged that a society had to support people with different perspectives; however, he did not explain how people could develop an imaginative capacity beyond warning teachers not to suppress it when it appeared.[2]

Bateson showed how metaphors provided the means to imagine what might happen under conditions different from what existed at the moment. An example of Bateson's use of metaphors was his development of the term "schismogenesis." When Bateson had finished his field work with the Iatmul, he returned to Britain shortly before World War II

DOI: 10.1057/9781137484215.0007

to write his account of the observations he had made in New Guinea. Speaking with a friend about the ways the interactions among European countries were growing increasingly aggressive, he realized that the men and the women of the Iatmul responded to each other competitively in the same way nations engaged in an arms race. He mixed this metaphor with the dialectic from Hegel and constructed the theory of schismogenesis. This was the tendencies of people to direct their actions according to the responses of other people.[3]

For Bateson, metaphors did more than inspire sociological concepts. They offered ways to realize connections among actions that people tended to overlook. This appeared in his concern for environmental destruction as people tried to solve simple problems, such as eliminating mosquitoes with powerful insecticides. The resulting pollution aggravated the problems.

By presenting the work of the philosophers of education, this book suggests that teachers can approach the problems of environmental destruction by seeking traditional instructional aims. The point of this book is that understanding is an essential aspect of any plan for action. As noted in the introduction, the solutions are easy. Cleaning up a street or protesting efforts to weaken environmental regulations can appear as ways to introduce students to environmental concerns. The problem is that these actions may not enable students to recognize the complications in applying an ethical framework consistently. This will be explained in this chapter.

Some commentators claim the problems of environmental destruction are so serious and pressing that immediate and drastic action to change society is necessary. For example, Naomi Klein quoted scientists who warned that global capitalism had depleted the resources on the earth to such an extent that the process threatened to extinguish humanity. In her book *This Changes Everything: Capitalism vs The Climate*, she contended that the best hope for saving humanity was through mass protest movements that could change the ways capitalism abused the environment and the living things in it. Seeking examples of such protests changing society, Klein pointed to the U.S. civil rights movement, the second wave feminist movement, and the pressure for gay and lesbian rights. These movements had limited success. The most successful was the abolitionist movement in the United States during the nineteenth century. It ended slavery and the economic system on which it depended. Acknowledging that fossil fuel divestment and local laws prohibiting what she called high

DOI: 10.1057/9781137484215.0007

risk extraction of oil and gas were modest, she believed these movements might grow into something that could force multi-national corporations to forfeit future earnings so that forests and oceans might be saved. Her hope was that the reality of total destruction caused by climate change would be enough to mobilize protest movements to force the end of a capitalist economic system where profits came before life itself.[4]

According to Klein, the economic system had to change. She noted that the civil rights movement and the feminist movement won in court rooms, but the victories left the economic system that profited from human exploitation unchanged. Economic relationships seemed to be too large and too complicated to reform; however, she added that the massive reforms required to remedy the threats of climate change altered everything. For this reason, those threats could fulfill the promises of those protests. The movement for ecological preservation could unite those protests that remained active and correct the problems that previous protests attacked in isolation because the efforts to end climate change required a transformation in people's world views.[5]

Klein offered four related strategies to bring about what she called "deep social change." First, she recommended campaigning for things such as a guaranteed minimum income rather than for imposing a carbon tax. The guaranteed income might enable workers to refuse jobs that contributed to pollution. More important, the debate could provide opportunities to discuss the relative values of people helping each other and of protecting corporate profits. Second, these efforts could align with efforts to change the traditional descriptions of humankind in which stories show people as caring instead of greedy. Third, she urged people to encourage the diffusion of a worldview built upon a vision of humanity as interdependent rather than individualistic, caring rather than competitive, cooperative rather than domineering. Fourth, in all discussions about climate change, she wanted the emphasis to fall on moral values rather than on pecuniary ones. Although the costs of enslaving human beings outweighed the benefits, Klein claimed that abolitionists did not end slavery by showing these added expenses. She thought the victory came from depicting the slaveholders as ruthless, crude, and barbaric, which undercut the moral justifications of slave holding.[6]

In fairness, since Klein worked as a journalist, her recommendations fit the aims of journalism. For example, in describing the predictions scientists made of environmental destruction, journalists might feel that they could prevent the catastrophe by spreading the information

DOI: 10.1057/9781137484215.0007

that would inspire social reform. Unfortunately, social scientists have suggested that such warnings would not be enough to dismantle the entrenched economic system.

Writing in 2000, Seymour Martin Lipset and Gary Marks suggested that it would not be easy to change the economic system in the United States. The issue on which Lipset and Marks focused was the failure of socialism to become a reasonable political alternative as industrialism grew. To determine what made capitalism so impervious to reform, they pointed to many socialist authors who described America as exceptional. Those authors built on Tocqueville's analysis that America was unlike other countries in Europe or Asia. It had a high rate of equality, a productive economy, and the absence of feudal traditions. Holding to an ideal of individualism, Americans rejected the communitarianism associated with socialism and with environmentalism, and they distrusted the idea of forming a strong state to reduce inequality and to prevent pollution. Lipset and Marks noted that these tendencies made it difficult for either socialism or a Green party to become popular despite the hardships that capitalism would inflict on the voters. These tendencies distinguished America from other countries in Europe where socialism and a Green party had made some inroads.[7]

Klein hoped that deep social change would arise from a shift in world views. By this, she meant a shift in people's moral perspectives. Among the moral shifts that Klein advocated was the view of people as competitive. She thought this should become a perception of people as caring and cooperative; however, Lipset and Marks contended that Americans had retained a capitalistic orientation despite the harm it caused. Evidence supporting the contentions of Lipset and Marks appeared in several suggestions to remedy environmental destruction. The proposals retained the capitalist model and blamed the problems on people who were economically deprived.

An example of preserving the environment by harming the surroundings of economically deprived people appeared in the United States when several cities allowed private conservancies to create and to maintain urban parks. Critics complained that this practice resulted in selfish rather than environmentally sound practices. For example, David Callahan contended that wealthy philanthropists donated funds to create parks in New York City, Philadelphia, Houston, and Tulsa. The process began in the 1980s, Callahan explained, when wealthy private donors organized the Central Park Conservancy. Other privately funded

DOI: 10.1057/9781137484215.0007

organizations followed, such as Friends of Hudson River Park. While these groups created and maintained beautiful open spaces, those parks were near the apartments or office buildings of the wealthy donors. The parks in other less affluent neighborhoods declined because the public sector was burdened with budget cuts, and officials considered the donations of the privately funded organizations as maintaining the budget for parks which would have been reduced otherwise. According to Callahan, the process was aided by the growing inequality of wealth in the United States where the net worth of the people listed in the Forbes 400 had doubled from 2009 to 2014.[8]

Further support of the view of Lipset and Marks came from people who used the issue of declining green space to oppose the reform of immigration policies into the United States. For example, in a publication entitled *Vanishing Open Spaces*, the authors for NumbersUSA argued that "nearly all long-term population growth is related to federal immigration policies that have increased the annual settlement of immigrants from one-quarter of a million in the 1950s to more than a million a year since 1990." The report added that the only way to end the destruction of farmlands and forests was to reduce the levels of immigration.[9]

This publication was not an isolated complaint from NumbersUSA. Writing in 2014, Julie Hirschfeld Davis claimed that the organization NumbersUSA was a powerful voice that had helped to stop every effort at immigration reform in the United States for two decades. She added that when she wrote her article, NumbersUSA employed only 35 staff members and had an annual budget of $10 million; however, it had overcome the efforts of a coalition of groups that spent about $1.5 billion from 2008 to 2012 in support of immigration reform.[10]

To explain the link between the growth of the human population and ecological destruction, NumbersUSA advertised on its web site a collection of essays written by many environmentalists who shared this view. Entitled *Life on the Brink*, these essays asserted that overpopulation was the main issue concerning ecological preservation. The authors included professors from the University of Colorado, Washington State University, Stanford University, University of Denver, and the University of California at Santa Barbara.[11]

Philip Cafaro was one of the editors of *Life on the Brink*. His blog appeared on the NumbersUSA website. For his edited volume, Cafaro wrote an essay with Winthrop Staples III that presented the argument found in *Vanishing Open Spaces* in a syllogistic form. It consisted of six

DOI: 10.1057/9781137484215.0007

premises that led inexorably to a policy that would end immigration into the United States and reduce the population in the world. Those premises asserted that immigration drove population growth in the United States, and this created environmental problems. The increased population used more land and added to the pressure on other parts of the world to support the high standard of living in the United States, which contributed to environmental destruction in other countries. These premises led Cafaro and Staples to a policy with three steps. First, Americans should reduce their consumption of goods. Second, Americans had to eliminate illegal immigration to lock in any gains in their own country made by such reductions. Third, the American government should provide foreign aid only to those countries that provide contraception and abortion to their citizens. The reason for the restrictions on foreign loans was to encourage all countries to reduce the world's population and the pressure on immigration to the United States.[12]

The ideas behind the private conservancies and of NumbersUSA represent an approach to conservation that expands the problems Tocqueville described. Based on selfish disregard for the democratic notion of shared goods, the private conservancies and NumbersUSA suggest solutions that would make the problems worse. As described earlier, Bateson showed that the problems of pollution derive from the myth of power. When people believe they can solve their problems by changing the environment, this misperception begins a chain of events that increase the difficulties. Is not difficult to see that labeling economically deprived people as unworthy of natural beauty will make them contemptuous of the beautiful surroundings wealthy people have for themselves. Such attitudes cannot reinforce environmentally friendly actions.

In a simple way, private conservancies and NumbersUSA show the attitudes Tocqueville warned against remain strong in the United States. More careful evidence comes from social scientists. For example, in 1985, Robert Bellah and his coauthors analyzed American society to determine if Tocqueville's complaint that social connections could not withstand the corrosive nature of individualism and its accompanying feeling of equality of condition that were endemic to American democracy.

Bellah and his coauthors interviewed more than 200 white, middle class Americans to determine how these people made sense of their lives, how they thought about their relation to society, and how they related their ideas to their actions. Although the people they interviewed held different things to be important, they seemed to agree that material

DOI: 10.1057/9781137484215.0007

success made other things possible. For example, one subject thought his family life was central to his existence, but he was driven to be a successful businessperson to make the family comfortable. Although Bellah and his coauthors found that the dominant value in contemporary culture was individualism, they expressed optimism that a movement similar to the civil rights movement could introduce some sense of social cohesion. They added that the reinvestment of intrinsic rewards in work would have to be a central aspect of such a reform. The notion of intrinsic rewards as part of social reforms derived from Bellah's conclusion that most people worked for extrinsic, material rewards, such as salaries that they used for other desires, and denied the genuine and social worth of labor. Tocqueville had arrived at a similar conclusion about people's feelings about work in America.[13]

As noted earlier, some educators suggest that schools should organize students to engage in direct social protests to end climate change. They follow a popular model developed by Paulo Freire called "problem posing education." Freire built on the idea of praxis, which he defined as theory and practice leading to social change. The idea was that students would consider a problem, suggest a solution, apply the solution in some way, and return to the classroom to discuss the results. Freire thought that such a pattern would enable students to see the problems more clearly and recognize the available solutions.[14]

This book does not recommend such a practical method. The lesson the three philosophers offered was that the important element for social change was not for protestors to follow some particular strategy. It was improving the ways people thought. As noted earlier, Harris, Dewey, and Bateson followed ideas of Hegel in different degrees and in somewhat different directions. To some extent, they built on each other's conceptions. While Harris borrowed his conception of psychology from Hegel, Dewey added the ideas of William James. Bateson may have mixed Hegel in his thinking, but he seemed to follow Bertrand Russell, who gave him the theory of logical types, described in an earlier chapter. Russell claimed that almost all of Hegel's ideas were false.[15] It may be that the way Bateson followed Hegel's lead was by integrating several views of philosophers who had worked previously into a coherent picture of human thought.

At any rate, the lesson is that recognizing ways to improve one's thinking is an important step in reducing environmental devastation. This is especially true for educators who could pass those lessons on to their

DOI: 10.1057/9781137484215.0007

students. If Bateson was correct in thinking that taking direct action to change the environment worsened the problems, the best approach to environmentalism would be to consider the epistemological errors in one's own thinking before charging ahead in some effort to reform society.

To aid in such reconstruction of perspectives, this book describes how the intellectual traditions within American culture provide a foundation for environmentalism. This may appear to be ironic because the premise of the book is that three aspects Tocqueville found in American democracy advanced such destruction. Fortunately, the irony is more apparent than real. Further, correcting the errors within the ideas that led people astray may offer more hope than it would to import suggestions from European or Asian thinkers who would make diagnoses from afar. For these reasons, the book looks for the basis of a reform from American philosophers: Harris, Dewey, and Bateson.

Bateson might appear to be an exception because he was born and educated in Britain. Nonetheless, he became a U.S. citizen in 1956, enjoyed a career in this country as a prolific author, and from 1976 to 1980, he served on the University of California's Board of Regents. A description of a letter he wrote criticizing the university's curriculum, entitled "Time is Out of Joint," appears in the chapter on Bateson.

The advantage of relying on American intellectuals is that they spoke from the same cultural surroundings that they analyzed. In this way, they offered suggestions that should be appropriate to the conditions in which Americans live.

There is no question that climate change and environmental pollution represent enormous problems. For example, in 2015, a group of researchers estimated that in 2010 about 12 million metric tons of mismanaged or untreated plastic waste went into the ocean. Such an annual rate of disposal was unsustainable because the dumped plastic reappeared in Arctic sea ice, on the sea floor, and in floating masses on the surface. It was eaten by sea life and posed dangers to people and other animals. Despite the dangers, the authors noted that the flow of plastic waste into the oceans continued unabated.[16]

Despite the pressing nature of the problem, the ideas this book has presented suggest that people can save themselves if they learn to avoid the dangers Tocqueville noted. Otherwise, the tendencies of individualism, materialism, and conformity might destroy democracy, the environment, and human life.

DOI: 10.1057/9781137484215.0007

Notes

1 John Dewey, *The School and Society & The Child and the Curriculum* (1915, repr., Mineola, NY: Dover Publications, 2001), 32–33.

2 John Dewey, *Democracy and Education: An Introduction to the Philosophy of Education* (1919, repr., New York: Simon & Schuster, 1944), 302–305.

3 Gregory Bateson, *Naven: A Survey of the Problems suggested by the Composite Picture of the Culture of a New Guinea Tribe drawn from Three Points of View*, 2nd edition (Stanford, CA: Stanford University Press, 1958), 260–261.

4 Naomi Klein, *This Changes Everything: Capitalism vs. The Climate* (New York: Simon and Schuster, 2014), 449–466.

5 Ibid.

6 Ibid.

7 Seymour Martin Lipset and Gary Marks, *It Didn't Happen Here: Why Socialism Failed in the United States* (New York: W.W. Norton, 2000), 15–41, 293, 294.

8 David Callahan, "The Billionaires' Park," Op-Ed, *New York Times*, Monday, 1 December 2014, p. A25; For a similar view, see Felix Salmon, "Why Privately-Financed Public Parks Are a Bad Idea," Reuters edition, U.S., 22 November 2013, http://blogs.reuters.com/felix-salmon/2013/11/21/why-privately-financed-public-parks-are-a-bad-idea/, accessed 6 December 2014.

9 Leon Kolankiewicz, Roy Beck, and Anne Manetas, *Vanishing Open Spaces: Population Growth and Sprawl in America*, Paper presented at Earth Day Texas Eco Expo, 26–27 April 2014 (NumbersUSA), xi, https://www.numbersusa.com/problems/environmental-impact, accessed 13 December 2014.

10 Julie Hirschfeld Davis, "Genial Force behind Bitter Opposition to Overhaul," *New York Times* National Edition, vol. CLVIV, No. 56, 705, Thursday, 4 December 2014, A20.

11 Philip Cafaro and Eileen Crist, eds., *Life on the Brink: Environmentalists Confront Overpopulation* (Athens: University of Georgia Press, 2012).

12 Philip Cafaro and Winthrop Staples III, "The Environmental Argument for Reducing Immigration into the United States," in *Life on the Brink: Environmentalists Confront Overpopulation*, eds. Philip Cafaro and Eileen Crist (Athens: University of Georgia Press, 2012), 172–188.

13 Robert Bellah, Richard Madsen, William M. Sullivan, Ann Swidler, and Stephen Tipton, *Habits of the Heart: Individualism and Commitment in American Life* (New York: Harper and Row, 1985), vi–xii, 20–26, 275–286.

14 Paulo Freire, *Pedagogy of the Oppressed*, trans. Myra Bergman Ramos (New York: Herder and Herder, 1970).

15 Bertrand Russell, *A History of Western Philosophy* (New York: Simon and Schuster, 1945), 730.

16 Jenna R. Jambeck et al. "Plastic Waste Inputs from Land into the Ocean," *Science*, vol. 347, no. 6223, 768–771, download: http://www.sciencemag.org, accessed 12 February 2015.

DOI: 10.1057/9781137484215.0007

Index

DOI: 10.1057/9781137484215.0008

DOI: 10.1057/9781137484215.0008

DOI: 10.1057/9781137484215.0008

DOI: 10.1057/9781137484215.0008

CPSIA information can be obtained at www.ICGtesting.com
Printed in the USA
LVOW10*1156090915

453313LV00004B/4/P

9 781137 484208